CRITICAL ANIMAL STUDIES

CRITICAL ANIMAL STUDIES

AN INTRODUCTION

Dawne McCance

SUNY PRESS

Published by State University of New York Press, Albany

For information, contact State University of New York Press, Albany, NY
www.sunypress.edu

Production by Diane Ganeles
Marketing by Anne M. Valentine

Library of Congress Cataloging-in-Publication Data

McCance, Dawne, 1944–
 Critical animal studies : an introduction / Dawne McCance.
    p. cm.
 Includes bibliographical references (p.   ) and index.
 ISBN 978-1-4384-4535-9 (alk. paper) — ISBN 978-1-4384-4534-2 (pbk. :
alk. paper)
 1. Animal rights. 2. Animal welfare—Moral and ethical aspects.
I. Title.
 HV4708.M387 2013
 179'.3—dc23
                                                        2012009562

10 9 8 7 6 5 4 3 2 1

This book is for Erin

Contents

Preface

Chopped Liver

My first experience as a graduate student, newly admitted to a master of science program in biochemistry, involved protein synthesis experimentation conducted on rats in a medical science laboratory. Actually, the basement room of the since-condemned and demolished building in which most of this research took place bore little resemblance to the university labs I had imagined or previously encountered. The entire research facility was old and crumbling, the basement area particularly bleak. The "animal lab" to which my supervisor assigned me—a windowless and dimly lighted cell—contained sixty wire mesh cages suspended in rows on a metal frame. A table against one wall held an old, and very dull, guillotine—smaller than a paper cutter, and not nearly as efficient. My assignment as a student researcher was to nurture a new shipment of rats (they arrived soon after I did), that is, to feed them carefully allotted daily portions of a diet formulated to determine the ingredients essential to protein synthesis. I was instructed to weigh the rats daily, and when they achieved the desired mass, to decapitate them, cut open the thorax and remove the liver, and take it to a secondary lab upstairs. There, usually in late afternoon sunlight after my supervisor had left for the day, I would queasily chop the still-warm liver into small pieces, cauterize it in a centrifuge, and do a protein analysis on the results.

In the critical animal studies literature, the rodents favored for laboratory experimentation are not always counted as animals having a strong moral claim. For many, then, my rats would have had lower ontological and moral status than the puppies in the basement lab down the hall from mine, the lab that my student researcher colleagues called their "puppy mill." Although I did not give my rats names, such as those Jane Goodall bestowed on her Gombie chimpanzees, one thing I learned in working with them is that, as much as any puppies I have known, and probably much like chimpanzees, each has a "personality," temperament, and behavioral features all its own, features I recorded at the time in my journal. Marc Bekoff and Jessica Pierce (28), reporting on the work of neurobiologist Jack Panksepp, suggest that rats are social, experience joy, and laugh when tickled. I can assert that they squeal and scream in pain. They also demonstrate intelligence—the sort of intelligence a rat needs to determine how one's own cage might be pried open, and in turn the cages of one's fellow rats. When only five of the animals had been brought to the guillotine, this was the kind of "rat intelligence" that, somewhat to my relief but much to my supervisor's chagrin, put an end to the protein synthesis experiment: an overnight rebellion of rats releasing rats, and large rats eating small ones. The event brought not only a catastrophic halt to that summer's research, but also, with my decision to forego a career in experimental science, the close of my study of biochemistry. I remember my father's dismay at the news that I was withdrawing from the biochemistry graduate program, and my feeble answer to his question why: "If I study and teach philosophy, I won't have to spend my life cutting off heads."

At the time, I was clearly unaware of the nonliteral "putting to death," in which, as Jacques Derrida observes, philosophy and other humanist discourses participate "*to the extent that they do not sacrifice sacrifice*" (Derrida "Eating," 113). Is it fair to say that a "sacrificial structure" (113) prevails in today's critical animal studies and animal ethics, where one might least expect it to be found? It is a key question with which to open this book—and the question on which the book concludes.

Acknowledgments

I wish to acknowledge the support of the Social Sciences and Research Council of Canada, St. John's College, and the Department of Religion at the University of Manitoba. Thank you to David Farrell Krell for his support, including his reading of, and interest in, this book. Thank you to my graduate student, Bruce Conan, for proofreading the original manuscript. And thanks to Andrew Kenyon, Diane Ganeles, and the excellent staff at SUNY Press.

1

INTRODUCTION

Never mind that the astronomy of Nicolas Copernicus amounted, in Arthur Koestler's words (179), to no more than a ramshackle hodge-podge of epicycles, within fifty years of the 1543 publication of *De revolutionibus orbium coelestium* (*On the Revolutions of the Celestial Spheres*), the transition to mechanical philosophy, and thus to modernity, was underway. By the beginning of the seventeenth century, "the entire episteme of Western culture found its fundamental arrangement modified," Michel Foucault writes in *The Order of Things*; "the seventeenth century marks the disappearance of the old superstitious and magical beliefs and the entry of nature, at long last, into the scientific order" (54). A world previously understood "as a complex of kinships, resemblances, and affinities, and in which language and things were endlessly interwoven," gradually gave way to a configuration for which *analysis*, division into the smallest constituent units, is the fundamental way of knowing, and for which the activity of the mind consists no longer "in *drawing things together*, in setting out on a quest for everything that might reveal some sort of kinship, attraction, or secretly shared nature within them, but, on the contrary, in *discriminating*," dividing, separating identities from differences (54–55). Under this new mode of analysis, the word *individual* lost the meaning it had in medieval argument—that which is *indivisible*, comprehensible only as a whole and indivisible from the whole to which it belongs (Raymond Williams 161–165)—in favor of its modern connotation: a single, detached, and soon-autonomous entity, itself divisible into lower and higher parts, animal body and animating mind. As will emerge in the following chapters, it is animals—countless and nameless animals—who

1

continue to bear the burden of this modern bifurcation of the living from the dead.

Even before *De revolutionibus* was put on the Papal Index in 1616, Galileo set the stage for the bifurcation when, peering through his telescope at the moon, he saw, not the heavenly bodies of the ancients, but a barren, *dead*, lunar landscape. In his 1610 *Sidereus nuncius* (*Starry Messenger*), Galileo provided the seventeenth century with what Timothy Reiss suggests is its "most eloquent metaphor": *I/eye-instrument-world* or *mind-sign-world* (24). For one thing, the metaphor signals the birth of perspective, whereby the single, immobile I/eye that looks out upon the world (from the centric point of a visual pyramid or through a telescope) is both *disincarnated*, set apart from its body, and *detached*, set apart from what it sees (Jay *Downcast*, 70; "Scopic," 8; Panofsky *Perspective*, 27–36). As soon as this disincarnated I/eye claimed "clear and distinct" certainty for its lone viewpoint, the linguistic sign no longer retained its participatory sense, becoming instead as transparent as a telescope, not involved in constituting "truth" but serving merely as an instrument of it. With this withdrawal of language into "transparency and neutrality" (Foucault *Order*, 56), the new regime established itself as at once *analytical* and *referential* (see Reiss 24)—even if the "truth" of its "reifying male look," inevitably "turned its targets into stone" (Jay "Scopic," 8).

Most, if not all, contributors to critical animal studies would agree that, particularly since the seventeenth century, modern Western ways of knowing nonhuman animals, inseparable from violent techniques practiced on them, have turned animals into "stone," that is, into inert objects, useful and disposable things: *reproductive machines* is the term ethicist Peter Singer uses when discussing the fate of sows in today's industrialized hog farming, where the goal is to use all available manufacturing techniques to "produce" as many as possible pigs per sow per year, and to fast-track those pigs, those "products," to slaughter weight (Singer *Liberation*, 126). It is as if we have come full circle from the understanding and use of animals proposed in the seventeenth century by René Descartes (1596–1650), who, with his binary (two-term, either/or) mode of organizing all that exists, is often cast as the "father" of modern thought. Cartesian dualism, though it takes many forms, is rooted in a hierarchical intelligible/sensible, mind/body opposition: according to Descartes, the essence of the human lies in thought, the wholly immaterial mind, *res cogito*, which he declared to be entirely separate and detached from the material, bodily realm of *res extensa*. The latter he describes in his *Description of the*

Human Body and of all its Functions (all the while giving an account of the living body of the dog on which he is practicing vivisection) as nothing but a machine, the functions of which follow from the mere arrangement of its working parts (317). The line of division he introduces, it seems for all subsequent modernity, is between this *animal-automaton body*, as mere matter, extended substance, and *thought*, the incorporeal mind or soul, as pure interiority that, as he puts it in his *Discourse on the Method*, "does not require any place, or depend on any material thing" (27). Herein lies the difference, he says, between man and beast: while the human body, like the animal one, comprises nothing but a machine, the human alone also has a mind, is separable into *both* a rational, thinking being *and* an animal *bête-machine* (beast-machine).

Paradoxically enough, even in the burgeoning discourse of animal studies, Cartesian dualism still holds sway, having only recently come under critique from relative newcomers to the field. Indeed, in much of what we might call *formative* animal studies, the assumptions comprising the seventeenth-century meta-text remain more or less intact, for example: faith in the certainty and referentiality of "factual" knowledge, that of science, and no less, of the "science of ethics"; reduction of language, including the language of animal ethics, to a vehicle of referential truth; preponderance of the analytic method and of a fundamental distinction between body and mind; demotion of the biological-animal body and association of the essentially human with mind or mental capacity; reinstatement of the "I" (the self, person, or subject) as *alive* by virtue of its rational capacity, as *author* of ethics, and as the *norm* against which the moral worth of other living creatures is measured.

With regard to the latter point, feminist philosopher Kelly Oliver writes that while debates over the status of animals "have exploded, making a survey of the literature overwhelming," it remains the case that most of these debates "revolve around the ways in which animals are—or are not—like us, and therefore should—or should not—be treated like us" (Oliver *Lessons*, 25). This insidious "most like us" standard—which we will encounter often in the following pages, and which we will approach as speciesist, anthropocentric, subject-centered, and dualist at once—can also be traced to the seventeenth century and Descartes. In keeping with the analytic method, he broke down social groups into their composite building blocks, each unit or "single man" (*Discourse*, 116–119) a disembodied "I," like all other individuals in being an essentially rational mind (127). It remains a curiosity of the emerging individualism that it robbed this or that individual of specificity, reducing all to one and the same essential rationality, such that Descartes could point to *himself* as universal norm, representative of all rational men where it comes to distinguishing "the true from the false" (115). In animal studies, constitution of the ethicist as exemplum and ground of truth only repeats the Cartesian gesture, just as reducing animals, in all of their differences, to a single term, "the animal," perpetuates Descartes' representation of all animals as one and the same *bête-machine*, relegated to the other side of his fundamental mind/body divide.[1]

One purpose of this book is to introduce the field called "critical animal studies," which first emerged some forty years ago as a specialization within analytic philosophy, one that set out both to expose, and to offer ethical responses to, today's unprecedented subjection and exploitation of animals. In recent years, analytic philosophy has been joined by, although it has not always welcomed, other philosophical traditions and critical animal studies has expanded enormously to encompass many diverse academic and activist pursuits. My purpose, however, is not only to outline and describe these developments as broadly as one small book will allow, but also, in line with some more recent contributions to animal studies, to engage the field in a critical way—as the label *critical* animal studies invites.

I interpret the word *critical* here in three senses that are relevant to my task, and to the task of critical animal studies today (see Partridge 130). First (as derived from the Greek *krisis*, a sifting, *krinein*, to sift, and the adjective *kritikos*, able to discern), the word suggests concerned questioning of inherited conceptual frameworks and modes of action they inform, the kind of judgment or discernment that belongs to interpretation of the history of human exploitation of nonhuman animals and that extends to the thinking and practice of ethics today. This task includes questioning not only of the West's post-seventeenth-century legacy, but also of cultural and religious traditions that extend even into antiquity. Second, the word *critical* goes back to the Latin *criticus*, in grave condition, and *criticare*, to be extremely ill: given the side effects of today's mass mistreatment of animals—loss of biodiversity; extinction of species; pollution of water, air, and soil; antibiotic resistant diseases; global warming, and so on—this sense of critical as crisis cannot be lost on "critical animal studies." Nor are we likely to feel well, or to feel well about ourselves, under still-persisting Cartesian mind/body, man/animal dualities. But it is not too late, the sifting has begun, and this leads to a third meaning of critical that relates to the crisis or turning point of a disease, a hinge, a pivot point where things might just turn around and go another, and better, way (see "hinge" in the *Oxford English Dictionary*). I am interested in foregrounding these and other historically aware and self-critical attempts by men and women to hinge critical animal studies; that is, to rethink its cardinal conceptual supports, and thereby, enable it to turn.

2

Animal Liberation on the Factory Farm

In his Foreword to *Animal Others* (ed. Steeves), Tom Regan offers a brief account of the origin and development of today's discourse of animal studies, and a compelling account at that. In 1971, he notes, three young Oxford University philosophers, Roslind and Stanley Godlovitch and John Harris published *Animals, Men and Morals*, a book that "marked the first time philosophers had collaborated to craft a book that dealt with the moral status of nonhuman animals. At the time of its publication," Regan writes, "the editors could not have understood how important their effort would prove to be. Or why" (xi). Regan's story goes on: Peter Singer, at the time another young Oxford philosopher, was so impressed with the book that he sent an unsolicited review of it to the *New York Review of Books* (NYRB). The editors of the NYRB not only published Singer's review in 1973, but also asked him to consider writing such a book himself. Two years later, Regan writes, "Singer's *Animal Liberation* burst upon the scene. From that day forward, 'the animal question' had a place at the table set by Oxbridge-style analytic moral philosophers, and a legitimate place at that. In the past twenty-five years, these philosophers have written more on 'the animal question' than philosophers of whatever stripe had written in the previous two thousand" (xi). The contemporary animal studies movement was on its way.

Regan's account belongs to what has become something of a master-narrative in the field of animal studies, a narrative that traces significant philosophical concern with the moral status of nonhuman animals back

7

only to the 1970s and to "Oxbridge-style" analytic moral philosophy, specifically to Singer's *Animal Liberation*, and to the wealth of publications that Singer's book has generated from philosophers working in the Anglo-American tradition. Perhaps it is appropriate, then, that the present book introduces readers to critical animal studies by way of the work that Peter Singer has done in calling attention to, and documenting, issues involving cruelty to, and maltreatment of, nonhuman animals: provoking discussion and debate; promoting advocacy and activism as well as academic participation in the animal studies field; working collaboratively with others; and above all, bringing analytic ethics, particularly his own utilitarian perspective, to bear on many pressing animal welfare issues. Ira W. DeCamp, professor of bioethics in the University Center for Human Values at Princeton University and laureate professor in the Centre for Applied Philosophy and Public Ethics at the University of Melbourne, Singer is author of *Animal Liberation* (first published in 1975, revised and reissued in 1990, 2002, and 2009), *Practical Ethics* (first published in 1979 and reprinted several times), *Animal Factories* (1980, co-authored with Jim Mason), *The Ethics of What We Eat* (2006, co-authored with Jim Mason), *In Defense of Animals: The Second Wave* (2006, an edited anthology), and numerous other publications. As will emerge from this and subsequent chapters, his work has proven central in shaping animal studies, although as the field has developed, the sufficiency, even moral legitimacy, of his utilitarian "equal interests" approach has been called into serious question. Given the important place he gives it in *Animal Liberation*, factory farming serves in this chapter as our entry point to Singer's animal ethics.

What Is Factory Farming?

Matt Ball, co-founder of Vegan Outreach, offers a brief description of the fate of today's animals in farm factories: "layer hens with open sores, covered with feces, sharing their tiny cage with the decomposing corpses of fellow birds; pigs sodomized with metal poles, beaten with bricks, skinned while still conscious; steers, pigs, and birds desperately struggling on the slaughterhouse floor after their throats are cut" (182). No words can convey the horror of what goes on in these factories, Ball maintains, and not even videotapes can "communicate the smell, the noise, the desperation, and, most of all, the fact that each of these animals—and billions more unseen by any camera or any caring eye—continues to suffer like this, every minute of every day" (182). One of the decided benefits of Singer's *Animal Liberation* is its well-researched resource of information on specific aspects of these factory farm operations in the United States and around the world. Together with *Animal Factories* (1980), and *The Ethics of What We Eat* (2006), books Singer co-edited with Jim Mason, *Animal Liberation* provides succinct depictions, as graphic as Matt Ball's, of the large-scale, highly mechanized industry, the "modern animal factory," that since the 1960s has been rapidly replacing family farms.

In *Animal Factories*, Mason, who grew up on a typical American family farm in Missouri, suggests that on such farms animals "were truly domestic, that is, they were part of the farm household. Families lived from and cared for their animals. Through natural growth and reproduction, the animals produced enough to supply the household and bring in a small but steady flow of cash from sales in nearby markets" (xiii). We are now seeing the end of the centuries-old era of family farms.[1] The industrial operations that have overtaken them confine animals for most or all of their lives within densely crowded buildings, without access to sunlight, grass, or vegetation, and often without room to stand or turn around. "Animals are reared in huge buildings, crowded in with cages stacked up like so many shipping crates. On the factory farms there are no pastures, no streams, no seasons, not even day and night," Mason writes. "Animal-wise herdsmen and milkmaids have been replaced by automated feeders, computers, closed-circuit television, and vacuum pumps. Health and productivity come not from frolics in sunny meadows but from syringes and additive-laced food" (xiii).

Although *Animal Factories* was published in 1980, its description of the transformation of farm animals into industry-produced "biomachines" still applies well to the factory farms of today—save that these, monopolized now by fewer owners, are even bigger, more industrialized, and more indifferent to animal welfare than ever before. From the start, factory farming has involved brutality toward animals, something that was evident to Singer when he first published *Animal Liberation* in 1975. In chapter 3 of that book, "Down on the factory farm . . . *or what happened to your dinner when it was still an animal*," updated for the 2009 edition, Singer contends that today's "use and abuse of animals raised for food far exceeds, in sheer numbers of animals affected, any other kind of mistreatment. Over 100 million cows, pigs, and sheep are raised and slaughtered in the United States alone each year; and for poultry the figure is a staggering 5 billion," which means, he adds, that on our dinner table and in our supermarkets, "we are brought into direct touch with the most extensive exploitation of other species that has ever existed" (95).

Rather than "factory farms," the industry prefers to call its agri-business facilities CAFOs, Concentrated (or Confined) Animal Feeding Operations, or simply AFOs, Animal Feeding Operations. Far-removed from the family farms of yesteryear, these facilities are specialized, large-scale, intensive, fast-track, and corporate-owned enterprises that distribute their products (meat, milk, and eggs) to national and international markets. Invariably, they aggregate thousands of animals within crowded conditions, often in indoor facilities and, as designed-for-profit operations, they use assembly-line industrial methods, hormones, antibiotics, food additives, genetic and biotechnologies to produce the highest possible yield of the type of product that consumers desire. For example, by accelerating the metabolism of "broiler" chickens ("table" chickens raised to be eaten), factory farms produce larger, meatier, birds that require less food, grow rapidly, and so can be slaughtered after only a few weeks. Whereas the natural lifespan of a chicken is about seven years, factory farmed broiler chickens, bigger than ever before, are slaughtered at seven weeks, Singer notes in *Animal Liberation* (99). Typically, then, this kind of fast-track farming results in shorter and meaner lives for animals, but greater yield, and massive profit, for corporate owners. Specialized factory production of animals has overtaken dairy farming, the raising of cattle, sheep, pigs, geese, ducks, and even fish. What follows here, preliminary to an outline and analysis of Singer's ethical approach to factory farming, is a brief portrait of the kind of CAFO operations that have all but displaced family farms in the raising of chickens for eggs and meat, cattle for dairy products and meat, and hogs.

Chickens

Factory farming turned first to chickens and the industrialization of egg production. Whereas on traditional family farms, chickens usually ran free in barnyards, lived off the land by foraging, expressed their natural behaviors of moving freely, nest-building, dust-bathing, escaping from more aggressive animals, defecating away from their nests, and in general, "fulfilling their natures as chickens" (Rollin 11), the situation they experience on factory farms is vastly different—and well-documented and described in animal studies literature (see for example, Imhoff, ed., *The CAFO Reader*) and on animal welfare websites. Once hatched in industrial breeding houses, female egg-laying chicks are separated from males. They do not lay eggs, and because their flesh is of poor quality, male chicks "are, literally, thrown away. We watched at one hatchery as 'chick-pullers' weeded males from each tray and dropped them into heavy-duty plastic bags. Our guide explained: 'we put them in a bag and let them suffocate. A mink farmer picks them up and feeds them to his mink'" (Mason and Singer 5). In *Animal Liberation*, Singer notes that some male chicks "are ground up, while still alive, to be turned into food for their sisters. At least 160 million birds are gassed, suffocated, or die this way every year in the United States alone. Just how many suffer each particular fate is impossible to tell, because no records are kept: the growers think of getting rid of male chicks as we think of putting out the trash" (108).

Female chicks are transferred to huge layer houses into which thousands of birds are crowded. The birds are confined in small wire cages that are often stacked in layers, each cage holding up to six or eight hens at a time, standing on top of each other with no room for any of them to stretch their wings. Laying hens spend their entire lives in these battery cages, incurring various foot and body injuries from the unforgiving wire floor and walls, defecating through the mesh to a collecting trench below, their food and water delivered by one conveyer system, their eggs collected by another. Because, when confined to so small a space they cannot establish a dominance hierarchy or pecking order, the birds are routinely "debeaked," to prevent their cannibalizing of each other. In *Animal Liberation* (101), Singer notes that the practice of debeaking began in San Diego in the 1940s, and used to be performed with a blow-torch, the farmer literally burning off the upper beaks of the chickens. "A modified soldering iron soon replaced this crude technique, and today specially designed guillotinelike devices with hot blades are the preferred instrument. The infant chick's beak is inserted into the instrument, and the hot blade cuts off the end of it. The procedure is carried out quickly, about fifteen birds a minute. Such haste means that the temperature and sharpness of the blade can vary, resulting in sloppy cutting and serious injury to the bird" (101). In animal philosopher Bernard Rollin's words, "The animal is now an inexpensive cog in a machine, part of a factory, and the cheapest part at that, and thus totally expendable" (11). Factory-farmed hens are raised to overproduce eggs. But at the age of about a year and a half, when their ability to produce eggs diminishes due to the wear and tear of cage life, and when it is no longer profitable to house and feed them, "they are made into soup and other processed foods" (Mason and Singer 5).

Chickens raised for meat, both males and females, once debeaked, are housed in huge broiler buildings where they are raised, for a mere seven or eight weeks of life, not necessarily in cages, but still in intensely crowded, unnatural, dimly lighted, poorly ventilated, and ammonia-filled conditions that allow each bird less than a square foot of space. As Singer points out in *Animal Liberation*, aside from stress and threat of suffocation, chickens raised in such broiler houses face many disease and health hazards, including sudden or "acute death syndrome," ulcerated feet, crippling and deformities, lung damage from inhalation of dust and ammonia, breast blisters, and hock burns (103–105). Like layer hens, factory-made broilers never see the light of day—at least until they are ushered from the semidarkness of the broiler building, loaded into crates, and piled into the back of a truck to be driven to a processing plant for slaughter. Such intense aggregation of birds in confinement conditions typifies the density that is characteristic of factory farming overall, and that is designed to produce the highest possible yield at the lowest possible cost. Understandably, under such conditions, antibiotics are regularly used to prevent costly disease outbreaks. Singer asks whether, once factory broilers are hung upside down and killed, plucked, dressed, and sold to millions of families who gnaw on their bones, people pause "for an instant to think that they are eating the dead body of a once living creature, or to ask what was done to that creature in order to enable them to buy and eat its body" (105).

The Dairy Industry

used to be 25 years
3–4 years of life now.

Factory-farmed dairy cows are usually confined indoors, in huge metal-roofed holding sheds or milking buildings, where they may be chained by the neck and/or cramped within individual stalls, their food and water delivered by conveyer-belt methods, their milking also mechanized; or else they are trough-fed in crowded and unsheltered outdoor "dry-lot" enclosures, that is, confined areas without grass, the floors of which are coated with urine and manure (a surface that leads to foot rot in many cattle). Industrialized dairy herds are enormous in size, comprising

hundreds or thousands of cows (Cross), none of whom ever see pasture. "The most obvious aspect of industrialized dry-lot dairying is the
virtual exclusion of that part of the farm land most closely associated
with the dairy farm, pasture," Howard F. Gregor writes in a 1963 study,
by which time in the United States, the shift from family dairy to factory farm was already well on its way (299). Another striking aspect to
which Gregor points, already evident by 1963, is the extremely large
herd sizes on industrial dairy farms: huge herds, high cattle densities, and
an unprecedented output of milk per number of cows (301). "Today's
dairy cow produces three to four times more milk than 60 years ago. In
1957, the average dairy cow produced between 500 and 600 pounds of
milk post lactation. Fifty years later, it is close to 20,000 pounds," Bernard
Rollin notes, citing American statistics from 2005 to 2006: "From 1995
to 2004 alone, milk production per cow increased 16%. The result is a
milkbag on legs, and unstable legs at that. A high percentage of the US
dairy herd is chronically lame" (12). Lameness is not surprising given
the quantity of milk that industrialized dairy cows are bred to produce,
and given that they may spend their lives standing in a mix of urine and
rotting manure. Cows may also suffer tenderness and swelling from injec-
tions of BST, bovine somatotrophin, a genetically engineered growth
hormone that is widely used in the United States to further boost milk
production—and that can also increase problems with mastitis. The latter condition, not uncommon among cows that carry such inordinately
heavy milk loads, involves infection and swelling of udders, which in turn
prompts widespread use of antibiotics within the industry and/or the
practice of tail amputation or docking (without anesthetic) to minimize
teat contamination with manure (Rollin 13). Dairy cows do not produce
milk until they have calved, so they are made pregnant as soon as they are
mature, usually through artificial insemination, and, adding to the misery
of their short lives (not twenty-five years, as in the past, but more like
three or four), industrialized dairy cows are separated from their calves
within hours of birth (so as to harvest the cow's milk for sale). "Dairy
cows usually give birth about three times in their lives, and every one of
their calves is taken away within 48 hours of birth," Erik Marcus writes
in *Meat Market*; "I think the trauma caused by these separations is one
reason why some cows are sent to slaughter early" (36).

Male calves, not useful for milk production, are often sold to the veal
industry, in which case they, too, live miserable—and very brief—lives,
tied in stalls and fed a milk-replacement diet that promotes rapid weight

gain, until at about 16 weeks they are slaughtered. For these male dairy calves, Mason and Singer suggest in *Animal Factories*, immediate slaughter is the better fate than those 16 weeks of confinement in semi-darkness, tied at the neck, in a crate too narrow to turn around (58). Indeed, the intensive farming of veal calves, although it does not compare in size with poultry, beef, or pig production, "ranks as the most morally repugnant" type of factory farming now practiced, Singer contends in *Animal Liberation* (129). He points to the discomfort caused by the slatted wooden floors of the stalls in which the calves are kept; the cruelty of removing them from their mothers and of frustrating their need to suck, and later to ruminate; the insidious practices of making the calves anemic, and therefore unhealthy, just so that their flesh will appear pale pink, and of depriving them of water so that they will take in more food (achieving the greatest possible weight in the shortest possible time); and keeping the veal sheds dark so as to reduce restlessness (129–136).

Beef Cattle

Beef cattle experience less confinement today than chickens, dairy cows, and hogs, and as Erik Marcus observes, beef calves are both born with the healthiest bodies of any factory farm animals and allowed a good quality of life: they grow up alongside their mothers and are allowed to graze in nature—at least for six months (39). The grazing time is short, however, relative to what it was twenty years ago, when calves roamed for about two years (Singer *Liberation*, 139). Now, at six months, they are forcefully removed from their mothers and shipped, sometimes long distances, to feedlots to be brought to slaughter weight. Prior to shipping, the calves are branded, dehorned, and castrated, the latter two, particularly painful, procedures performed without anesthetic and in the "gratuitously cruel" ways that Marcus describes in *Meat Market* (41–43). Once beef cattle arrive at a feedlot, their good quality of life comes to an end. There, they are given high-calorie grain feed, mainly corn, along with hormone implants to accelerate growth, increase muscle mass, and "finish" them as quickly as possible, that is, bring them to a slaughter weight of about 1,000 pounds. "It's a remarkable system, really—except for one complication. Because cattle evolved on a diet of grass and brush, corn-based feed damages their livers and causes other health problems," Marcus notes. And he goes on: "Cattle suffer from more than an inappropriate diet. Conditions at feedlots are bleak, crowded, and filthy. Just like cows at outdoor dry-lot dairies, feedlot cattle walk and sleep atop a blackened layer of dirt and manure. A single large feedlot may contain several thousand cattle, and the stench is unbelievable. There have been a number of occasions when I have smelled a feedlot from several miles away" (44).

According to Singer in *Animal Liberation*, the growth of large feedlots has been the dominant trend in the cattle industry, with 70% of the 34 million cattle slaughtered in 1987 in the United States having come from such operations (139). One problem he identifies with this type of factory farming, concurring with Marcus, concerns the adverse effects that concentrated feedlot diets, designed to maximize weight gain, have on beef cattle (140). Another serious problem, he says, again in agreement with Marcus, is exposure to the elements, whether hot summer sun without shade or winter cold and snow, conditions under which calves are especially vulnerable (140). Yet within today's massive factory farm operations, as we have noted, an animal is viewed not as a living being so much as a commodity, "as soulless as copper or scrap iron," as Steve Bjerklie puts it (143). Feedlot practices prevail because of their low production cost relative to how quickly they bring animals to slaughter weight. "In beef, grass-fed cattle gain weight steadily but slowly compared with animals that are 'finished,' to use the industry's word for it, for the last three to four months of their lives in feedlots on fattening feed grains" (143). By the same token, across the industry overall, instead of caring for weak or vulnerable animals, factory farms cull them, and at a significant rate. "As in any other industry, if a machine or production material is substandard, it is simply replaced" (Imhoff, 74). *select*

Pigs

A Worldwatch Institute Study published in 2005 reports that global meat production has increased more than fivefold since 1950 and more than doubled since the 1970s, with pork accounting for most of this production (Nierenberg 9). A stunning increase, one that comes with the move to intensive pig farming, which, as in other types of factory farming, involves the partial or total confinement of animals in a cramped, alien, and stench-filled indoor environment, mutilation, separation of sows from offspring, and in general, the reduction of animals to machines. In the case of intensive hog farming, the sow "is being turned into a living reproductive machine," Singer contends (*Liberation*, 126), because the industry goal is to use all available manufacturing techniques to "produce" as many pigs per sow per year as possible, and to fast-track these offspring to slaughter weight. Once again, with industrialized pig farming Singer writes, "the producer's profits and the interests of the animals are in conflict" (129).

In *Animal Liberation*, Singer discusses certain hog "manufacturing" practices in which the conflict is particularly stark, practices that are now tackled across the critical animal studies literature. These include: housing of pregnant sows in gestation crates; early weaning of piglets so that, after delivering her litter, a sow can quickly be made pregnant again either by a boar or by artificial insemination; ever-increasing automation of the reproduction process, for example through mechanical nursing and technically induced super-ovulation; tail docking; castration of male pigs to improve the taste of their meat; intense confinement systems that rule out exercise ("calorie burning") in favor of more weight gain per pound of food consumed (119–129). In all cases, to recall Rollin's words cited earlier, these practices make it impossible for factory-farmed pigs to "fulfill their natures" as pigs. The point is made by Marian Stamp Dawkins in an essay that asks what good animal welfare might be. Put simply, Dawkins says, "animal welfare is about animals being healthy and having what they want" (74). Good welfare means that "animals are content because they are not desperately searching for something they do not have, and, equally, that they are not fearful of something they are trying to get away from and can't. 'Having what they want' thus includes having the positive elements needed to satisfy them and *not* having to put up with the fear-producing, anxiety-producing, boredom-producing elements that they want to escape from or avoid" (74).

But "[r]ight now, most pigs in American can't really be pigs," Singer and Mason point out in *The Ethics of What We Eat* (100). What do pigs want that is denied to them by the hog industry? Many descriptions can be found of the ways in which pigs fulfill their natures as pigs, but I refer you here to Singer's brief discussion in *Animal Liberation* of pigs' instinctive behavior patterns. To begin with, pigs are social animals and they form stable social groups, building communal nests, and using dumping areas well away from the nest. They are also active, giving much of the day to rooting around the edge of woodlands. When they are ready for birth, sows leave the communal nest and build their own nest on a suitable site, digging a hole and lining it with grass and twigs. They give birth

thin and narrow flat strip

there and remain in this nest with their piglets for about nine days, after which they rejoin the group (119–120). Confined as they are in factory farms, for the most part in crowded conditions indoors, without straw or other bedding material, without room to move, expected simply to eat and sleep, stand up and lie down, pigs cannot be pigs, the social animals they are. They suffer from boredom, frustration, and stress (porcine stress syndrome), and probably for this reason, they may engage in biting each other's tails—a problem that the industry handles by tail docking, a painful procedure performed without anesthetic.[2] Deprived of exercise, they are unhealthy as well as stressed, often with feet and legs deformed from standing on concrete or slatted floors, and/or with lungs damaged by all of the ammonia they inhale. Not the least, the housing of pregnant sows in gestation crates too small to allow the animal to stand up, turn or walk, is a practice deplored by animal advocates, Singer included. Sows remain in these gestation crates, or sow stalls, for all of their reproductive lives, save for the periods when they are transferred to equally confining, and virtually immobilizing, farrowing crates to give birth and briefly suckle their piglets. Singer notes, "in natural conditions the sow is a highly active animal, spending several hours a day finding food, eating, and exploring her environment" (*Liberation*, 127), so the locking of sows into immobilizing stalls can be described only as cruel.

compartment in a stable booth   *give birth to a litter of pigs*

Handling and Slaughter

Sadly, when it comes to the handling and slaughter of factory farm animals, the issues only multiply in number and degree of brutality. Although methods of slaughter vary from one species to another, certain animal welfare issues are common across the board. These include: shipping of slaughter-ready animals over long distances in cramped conditions without food or water, resulting in many deaths from heat or cold; cruel treatment of animals during truck loading and unloading, in holding pens, and in their movement from pen to slaughterhouse; inadequate stunning of animals in the slaughterhouse by electric shock or stun gun, so that many are hung upside down on conveyor belts and skinned or dismembered alive; the slaughtering of animals at such high speeds that the number of animals cut apart while alive can only increase. "In the early 1970s, the largest slaughterhouses killed about 170 cattle an hour," Erik Marcus writes in *Meat Market*. "Today, the fastest American slaughterhouses kill 400 cattle per hour on each line they operate. That means that every nine seconds, the pistol crew must shoot another animal between the eyes. At these speeds, mistakes are bound to occur. Compare the American situation to that of Europe, where cattle slaughterhouses run at only about sixty animals per hour" (47). The animal studies literature is replete with reports of live-animal slaughter, a practice that continues and worsens either because slaughterhouse supervisors refuse to reduce line speed and/or because government inspectors turn a blind eye to such conditions.

One of the most graphic, disturbing, and well-researched studies of American slaughterhouse practices in handling and killing cattle, hogs, poultry, and horses is Gail Eisnitz's *Slaughterhouse*. Peter Singer describes the book on its back cover as "a stomach-churning, damning indictment of the meat industry." About its author, he adds that, "Eisnitz is a brilliant investigator, writes superbly, and has the courage and persistence of someone who knows that she is right." And on the book's subject matter, he says this: "No longer can anyone believe that in the United States there is adequate inspection and control of slaughterhouses. As Eisnitz convincingly shows, the meat industry is indifferent to animal suffering, exploitation of its workers, and liable to produce a product that is riddled with dangerous bacteria." Indeed, one of the disturbing points Eisnitz makes in her book concerns the inadequacy of inspection and control of American slaughterhouses. Notwithstanding the Humane Slaughter Act, passed by the United States Congress in 1958 and broadened in 1978, staggering cruelty to animals runs rampant in the American slaughtering industry, in part, she says, because the United States Department of Agriculture, charged with enforcing the Humane Slaughter Act, is aligned with the meat industry and so largely indifferent to violations that regularly occur.[3]

Eisnitz offers firsthand accounts of a number of such violations, for example, those that occur on the "kill line" and involve the mutilation and skinning of thousands upon thousands of animals that remain alive and conscious after the stun—because the stun equipment is faulty or because the line moves too quickly to allow for an adequate stun every time. In either case, cruelty is built into the process. The oppressive process takes its toll on slaughterhouse workers, too, and puts them in physical danger of kicking and falling animals. Injuries are as high as the employee turnover rate. And, as Eisnitz documents, along with the physical injuries come demoralizing and brutalizing effects that are conducive

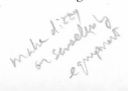

to alcohol abuse, family violence, and abusive behavior (see also Cook's "Sliced and Diced").[4]

One dimension of Eisnitz's study concerns the unwillingness of the American media, including network television stations, to make her findings known, purportedly for the reason that the story would prove too disturbing for the viewing or reading public. In her Afterword to the 2009 edition of *Slaughterhouse*, however, Eisnitz records her success in finally convincing the *Washington Post* to run a story, "They Die Piece by Piece," on one particular slaughterhouse at which numerous violations regularly occurred. The story, she says, "ended up being *one of the highest readership response pieces* in the history of the *Washington Post*" (299), and it prompted significant political response. Similarly, in Canada in June 2008, the national Canadian Broadcasting Corporation (CBC) aired a news investigation into the country's horse slaughter industry, which has grown rapidly since laws were passed in the United States making it illegal in that country to kill horses for food. To compile the CBC feature, reporters used hidden-camera footage from one slaughterhouse in Saskatchewan to document the cruel methods by which horses in that facility were being killed, and to make the case that, beyond that particular slaughterhouse, the business of killing horses for food (to be shipped to parts of Europe and Asia, where horse meat is considered a delicacy), operates both inhumanely and illegally. Public response to the CBC documentary was strong, and subsequent to that, the provincial Department of Agriculture announced its "commitment" to ensuring that in Saskatchewan, horses are slaughtered in humane ways.

Ill Effects of Fast-Track Farming

Sociologists note that the disappearance of the family farm, along with the small towns and rural communities they fostered, reflects the incursion of capital into the last domestic economic sector dominated by family businesses (Lobao and Meyer 118). It is not nostalgic to lament this passing, for as is becoming more evident every day, industrialized farming, with its litany of animal welfare issues, also raises numerous environmental and societal concerns. Among these, three particular concerns loom large: pollution, loss of biodiversity, and health issues.

Pollution is a major problem. Dairy, chicken, and pig factory farms produce enormous amounts of manure, in some cases, "as much excrement as a good-sized city" (Rollin 14). As one American study reports, on the North Carolina Coastal Plain alone, an estimated 124,000 metric tons of nitrogen and 29,000 metric tons of phosphorus are generated annually by livestock waste, which is either spread dry on fields or pumped into waste lagoons and then spread on fields as liquid, both disposal methods releasing large amounts of nitrogen and phosphorus into the environment (Mallin and Cahoon). With the huge quantities of manure they produce, factory farms fill the air with overwhelming odor that penetrates the homes and lungs of people living miles around. Carried in run-off water, manure and other wastes pollute rivers and streams, kill fish, lead to algae blooms, and upset aquatic ecosystems (see Tietz for an overview of hog production pollutants). And factory farm manure produces methane, a gas that contributes significantly to global warming (for a discussion of this, and an overview of the ill-effects of factory farming, and of arguments that dismiss these, see Debra Miller's *Factory Farming*).[5] As well, industrial pollutants, including factory farm pollutants, are now thought to be responsible for the shrinking bone density of Arctic people and animals, with the polar bears from east Greenland among the most polluted species.[6]

Loss of biodiversity is a consequence of the various contaminations mentioned above. Pollution of water systems degrades water quality and leads to habitats in which various fish and aquatic organisms cannot survive, virtually "killing" lakes and streams. Citing the Millennium Ecosystem Assessment Report published in 2005, commissioned by then–United Nations Secretary General Kofi Annan and representing the views of 1,350 scientists, environmental philosopher and lecturer Kate Rawles argues that large-scale farming increasingly disallows a diversity of other living things, and she notes that biodiversity loss (including the extinction of mammal, bird, and amphibian species) has been more rapid in the last 50 years than at any time in human history (47–48). Loss of biodiversity extends to wildlife as well. "Livestock now account for about 20% of the total animal biomass in the world and 30% of what they now occupy was once the habitat for wildlife," Marian Stamp Dawkins and Roland Bonney observe in their introduction to *The Future of Animal Farming*. "Through forest clearance, livestock farming could thus be said to be the biggest destroyer of biodiversity" (3). Biodiversity loss also results from industrial selection and its consequences in reducing "the numbers of livestock breeds in favor today and the essential genetic heritage they may be able to provide to present and future agriculturalists," and of course factory farming reduces "the number of farmers on the land" (Imhoff 161).

Health issues, for human and nonhuman animals, also follow from pollution and environmental degradation. We have already alluded to a number of the animal health and welfare issues involved in factory farming, including physical deformities, mutilations, infections, stress-induced disorders, production diseases (for example, liver diseases brought on by the high corn and grain diet cattle are fed to expedite "finishing"), and overall, short, miserable, painful (certainly not "healthy") lives—and brutal deaths. Disposal of the enormous amount of animal waste produced by industrialized farming operations poses health hazards for humans and animals. The Mallin and Cahoon study cited above explains that, unlike human waste, microbes (including pathogenic bacteria such as *Escherichia coli*, salmonella, and streptococcus, pathogenic protozoans, and a number of viruses) generated by factory farm wastes are not exposed to secondary treatment or chlorination to disinfect the material. Once dispersed on fields, ultraviolet radiation may deactivate the majority of these microbes, but given the sheer volume of waste distributed, there remains significant potential that pollutants will enter surface or ground water that humans contact, with particularly high dangers from waste lagoon spills (380–381). As we note above, these pollutants are poisoning water systems and killing aquatic animals.

Food safety concerns are also real, and have been exacerbated by panic associated with such outbreaks as the swine flu pandemic of 2009, the spread in 2001 of foot and mouth disease, and in the 1980s and 1990s of the so-called mad cow disease—bovine spongiform encephalopathy (BSE).[7] The intense crowding of animals in factory farms is conducive to pathogen spread, a point Eric Schlosser makes in *Fast Food Nation* (201). But as Danielle Nierenberg observes, it is really in the slaughterhouse that pathogens such as *E. coli* most efficiently spread, contaminating meat and leading to serious public health concerns. "Because animals' hides are covered in manure," she writes, "it's hard to keep fecal matter from coming in contact with the animal's flesh. In addition, when workers pull out the intestines of cattle, there is often what is called 'spillage'—literally the contents of the animal's digestive system spill everywhere. And modern slaughter and processing techniques frequently sacrifice food safety for speed: workers on high-velocity production lines may gut 60 or more animals an hour, making it easy for contamination to spread" (45–46). Slaughterhouse workers themselves face serious health risks, from a falling animal weighing one thousand pounds, from constant close contact with pathogens, from antibiotic resistant bacteria, from stress, and so on.

Even a brief overview of the ill-effects of factory farming suggests one reason why critical animal studies has seen exponential development in recent years, for as Marianne Dekoven argues (in "Why Animals Now?"), the turn toward animals is based, at least in part, on the idea that "the human species is destroying, and perhaps has irretrievably destroyed, the planet. There is no need to rehearse the evidence for this premise here—we know the facts about global warming; the already-existing alteration of normal weather patterns, promising continuous disaster; the disruption of ecological systems that support all life on earth; and the accelerating destruction of species and their habitats" (367). Dekoven surmises that many "have turned away from our own species in dismay at what it has wrought and turned toward other animals as a locus both of the other who calls us to ethics and of many of the things that, in our various modes of ethics, we value" (367). In what follows, we consider Peter Singer's utilitarian mode of ethics.

All Animals Are Equal?

Peter Singer titles the first chapter of *Animal Liberation* just this way: "All Animals Are Equal." The expression does not translate for him into the language of rights; it does not mean that all animals, human and nonhuman, have rights. The language of rights, he says, "is a convenient political shorthand" suited to "the era of thirty-second TV news clips," and "in no way necessary" in the argument for radical change in our attitude to animals (8). What does Peter Singer mean, then, when he proposes that "all animals are equal"? A first answer to this question might be that Singer rejects "speciesism." Coined by Oxford philosopher and psychologist Richard Ryder in the early 1970s, the term, as Singer defines it, "is a prejudice or attitude of bias in favor of the interests of members of one's own species and against those of members of other species" (Singer 2009, 6). In ruling out this prejudice, which he considers comparable to racism and sexism, Singer argues that membership in the species *Homo sapiens* is not a sufficient measure of moral worth, that species difference alone should not be allowed to determine the moral rightness or wrongness of a given action.[8]

This does not lead Singer to conclude that all nonhuman animals must be treated as equal to each other and to humans, any more than that all humans, despite their differences, must be treated in exactly the same way. Equality, he says, does not require equal treatment. To use the examples he gives in *Animal Liberation*, equality does not require that men, because they are equal to women, be granted the right to an abortion: "Since a man cannot have an abortion, it is meaningless to talk of his right to have one. Since dogs cannot vote, it is meaningless to talk of their right to vote" (2). What the principle of equality does require, Singer maintains, is "equal consideration of interests"—where the sole "interest" that he considers is that of avoiding suffering and of experiencing pleasure or enjoyment. On this key point, he refers to Jeremy Bentham's *Principles of Morals and Legislation* (1789), where, reflecting on the ill treatment of animals in eighteenth-century England, Bentham suggests that it is not "the faculty of reason" or the "faculty of discourse" (speech) that should be decisive where the moral worth is animals is concerned. "The question is not," Bentham writes, "Can they *reason*? nor, Can they *talk*?" but, "Can they *suffer*?" (qtd. in Singer *Liberation*, 7). According to Singer, this is the fundamental moral question: Can animals suffer? In order to have "interests," he argues, animals (and humans) must have the *capacity* to suffer. "The capacity for suffering and enjoyment is *a prerequisite for having any interests at all*, a condition that must be satisfied before we can speak of interests in a meaningful way" (7). A stone does not have interests because it cannot suffer, Singer explains, and thus it would be nonsense to say that it is not in the interests of a stone to be kicked along the road by a schoolboy. Stones cannot suffer, and therefore need not be taken into moral account. Animals can suffer, and therefore have interests that ethics cannot discount (7–8).[9]

However, Singer's answer to Bentham's "Can they suffer?" question is not a simple yes. For as evidenced in the above paragraph, Singer defines suffering as a *capacity*, one that is "not only necessary, but also sufficient for us to say that a being has interests" (*Liberation*, 8). As well, he ties the capacity for suffering to "mental capacities" or "mental powers," taking a "normal human adult," such as himself, as his moral benchmark (*Defense*, 5–6). One example he gives in his "Introduction" to *In Defense of Animals: The Second Wave* concerns a hypothetical decision to perform lethal scientific experiments on normal adult humans kidnapped at random from public parks. Soon, he suggests, every adult would be become fearful of entering a public park, and the resulting terror, along with loss of the enjoyment of visiting parks, "would be a form of suffering additional to whatever pain was involved in the experiments themselves." The same experiments performed on nonhuman animals would cause less suffering overall because "the animals would not have the same anticipatory dread."[10] In this case, his principle of equal consideration of interests would recommend using nonhuman animals, rather than humans, for the experiments, the reason being that "the superior mental powers of normal adult humans would make them suffer more" (5). In short, in Singer's answer to Bentham's question, "different mental capacities make a difference" (6).

As illustrated by the hypothetical case just mentioned, Singer's approach to ethics involves calculation or counting—where "like interests" (in avoiding suffering and experiencing pleasure) are counted equally—and as such, it represents so-called utilitarian theory. Writing in *In Defense of Animals: The Second Wave*, Gaverick Matheny defines utilitarianism as an ethical theory that takes as its rule, "'act in such a way as to maximize the expected satisfaction of interests in the world, equally considered'" (14). The rule says that, "I should sum up the interests of all the parties affected by all my possible actions and choose the action that results in the greatest net satisfaction of interests" (14). In Singer's approach, where the interest at stake is freedom from suffering in favor of enjoyment or pleasure, the rule requires that one choose the action resulting in the greatest net provision of pleasure and freedom from pain—for those with "like interests," that is, with the requisite mental capacities on which the capacity for suffering depends. Utilitarianism operates on a case-by-case basis, calculating net interests in each case. It is "consequentialist," then, "because it evaluates the rightness or wrongness of an action by that action's expected *consequences*: the degree to which an action satisfies interests" (15).

With his utilitarian approach to animal ethics, including issues raised by factory farming, Singer sees himself breaking with the Cartesian relegation of animals to the status of mere machines incapable of suffering. In *Animal Liberation*, he refers to Descartes as "the absolute nadir" of Western thought, the "last, most bizarre, and—for the animals—most painful outcome of Christian doctrines," particularly the doctrine of human dominion over animals (20). Targeting the Cartesian argument that animals experience neither pleasure nor pain, and that, deprived of immortal souls, they have no consciousness either, Singer draws a direct connection between Descartes' *bête machine* and factory farming's *biomachine,* a point we noted in the Introduction. And he is not alone in this. As Mary Midgley writes in her contribution to the anthology, *The Future of Animal Farming: Renewing the Ancient Contract* (eds. Dawkins and Bonney), Descartes "laid down the deadly doctrine" that animals are automata, thereby providing the Industrial Revolution with a dualist and mechanistic theory that could only encourage callousness towards animals (23–24). Even today, Singer suggests at one point in *Animal Liberation,* citing Ruth Harrison's *Animal Machines,* cruelty to animals is acknowledged in the farm industry only where it bears negatively on profitability (98). Challenging this attitude head-on, he maintains that if a being suffers, "there can be no moral justification for refusing to take that suffering into consideration. No matter what the nature of the being, the principle of equality requires that its suffering be counted equally with the like suffering—in so far as rough comparisons can be made—of any other being" (*Practical,* 50).

As for factory farming, Singer finds nothing to justify the misery it inflicts on animals from birth through to their slaughter. Industrialized farming methods stem from speciesism, he contends, and as he remarks in *Animal Liberation,* once we place nonhuman animals outside of our sphere of moral consideration, the outcome is predictable (97). Not only for reason of the suffering the industry visits on animals, but also for its environmental costs and hazards to human health, Singer rules against rearing animals for food on a large scale (160). One of his purposes in writing about intensive farming is to expose the cruelty it involves, something the general population may not be aware of when buying food in a store or restaurant. His accounts include activist, academic, and scientific perspectives, as well as results of consultations with farmers and with industry practitioners themselves.

In *The Ethics of What We Eat,* Singer and Jim Mason, by way of following three American families, make the case that our food choices have an impact on others, that they are a form of political action, and that they involve ethical decision making. The idea is that if consumers knew more about the origins of the food they purchase—where it came from and through what processes it was produced—they would be inclined to choose ethically. This book, too, is replete with graphic descriptions of the cruelty to animals involved in contemporary industrialized modes of rearing animals, for instance broilers ("Entering the Chicken Shed" 24–26), turkeys ("A Day in the Life of a Turkey Inseminator" 28–29), pigs ("A Pig's Life" 44; "Making Bacon" 49–54), dairy cattle ("Tracking

Down Jake's Milk" 55–60), and beef cattle ("The Beef Industry" 60–67). As in *Animal Liberation*, these descriptions are intended to support the argument that vegetarianism makes practical and ethical sense. At the very least, readers are encouraged to avoid eating animals farmed in industrial settings (thus in *Animal Liberation* [170], lamb is temporarily excused because little of it is presently intensively produced), and wherever possible, flesh should be replaced with plant foods; factory-farm eggs with free-range eggs (if these are unavailable, eggs should be avoided), and milk and cheese should be replaced with soymilk, tofu, or other plant foods (177). That said, Singer acknowledges that in our speciesist world, where it is not easy to adhere strictly to what is morally right, a "reasonable and defensible plan of action is to change your diet at a measured pace with what you feel comfortable" (176). Further to this, in his Foreword to *The Future of Animal Farming* he writes that, while he continues to see the rearing of animals for food as a manifestation of speciesism, he has come to realize that the chances of persuading the majority of meat-eaters to abandon all animal products are remote. "It therefore seems better to pursue a different strategy," he concedes. "The animal movement should continue to promote a cruelty-free vegan lifestyle, and to encourage those who are not vegans to eat less meat and dairy products. Recognizing that not everyone is ready to make such changes, however, the movement should also be involved in improving the welfare of animals used in commercial farming" (vii–viii).

These qualifications are pragmatic. In keeping with his "equal consideration of like interests" approach, Singer introduces some ethical qualifications as well. For instance, in *Animal Liberation*, he raises the problem of "line drawing"—a problem that turns on the question of whether a being has or does not have interests, that is, whether or not it has the capacity to suffer. "As we proceed down the evolutionary scale," he argues, the evidence for a capacity to feel pain (writhing, crying out, attempting to escape pain on the one hand; and on the other, possession of a nervous system similar to "our own") diminishes (171). On consideration of central nervous system evidence, he includes birds and mammals, reptiles, and fish as capable of feeling pain (171–172). Other forms of marine life, such as lobster, crabs, and shrimps "have nervous systems very different from our own," yet Singer would give them "the benefit of the doubt" (174). And so on: the point is that here, as elsewhere, "we" are the standard against which other beings are measured, and drawing the line between "higher" and "lower" depends on possession of nervous system ("mental") capacities like "our own."

Does Killing Matter?

As part of his analysis of industrialized farming in *Animal Liberation*, Singer includes descriptions of brutal slaughtering practices and of cruel methods of handling and transporting animals marked for slaughter. None of this suffering is necessary, he suggests. "Death, though never pleasant, need not be painful. If all goes according to plan, in developed nations with humane slaughter laws, death comes quickly and painlessly" (150). The implication is that killing animals is not so much at issue for him as is the painful practices by which killing is done. And indeed it is the case that, while the question of taking life is not absent from Singer's ethics, and while he considers the question complex, his approach to animal ethics focuses not on taking life, but on minimizing suffering. In *Animal Liberation*, he writes that the question of killing is one he prefers to "keep in the background" in favor of applying the "simple, straightforward principle of equal consideration of pain or pleasure," this as "a sufficient basis for identifying and protesting against all the major abuses of animals that human beings practice" (17). He does note, however, that in his nonspeciesist view, killing human beings is not always wrong, any more than is killing animals.

In illustrating this argument, as he does briefly both in *Animal Liberation* and *In Defense of Animals: The Second Wave*, Singer makes some startling suggestions, ones that any *critical* animal studies would need to assess with care. On more than one occasion, for example, arguing that species difference should not determine the wrongness of taking life, he proposes the following: "If it is wrong to take the life of a severely brain-damaged human infant, it must be at least as wrong to take the life of a dog or a pig at a comparable mental level" (*Defense*, 6; see also *Liberation*, 18–19). Or, what is more to his point: "perhaps it is *not* wrong to take the life of a severely brain-damaged human infant, at least when the parents agree that it is better that their child should die. After all, such infants are commonly 'allowed to die' in intensive-care units in major hospitals all over the world, and an infant who is 'allowed to die' ends up just as dead as one who is killed" (*Defense*, 6). Because the consequence is the same in either case—the infant "ends up dead"—painless killing, the action that is predicted to result in less suffering for infant and parents, is preferable for Singer.[11] This example leads to one of the questions that has been brought to his analysis of taking-of-life situations, as well as to his utilitarian approach overall: the question concerning his conception of "ethics," and whether ethics should involve more than a rote consequentialist calculation. According to Jacques Derrida, whose work we consider in later chapters, application of a preprogrammed calculus reduces ethics to a reaction, rather than a response—a point he makes while noting that the difference between response and reaction has been invoked by Western thinkers from René Descartes to Jacques Lacan in order to justify the man/animal binary (Derrida *The Animal*, 119–140). As we shall see throughout this book, the question of ethics, what it is and should be, is much at stake in critical animal studies.

Who Counts?

We need a new approach to the wrongness of killing, Singer argues, "one that considers the individual characteristics of the being whose life is at stake, rather than that being's species" (*Defense*, 6). In his view, such an approach would consider it more serious to kill "beings with the mental capacities of normal human adults" than to kill beings "who do not possess, and never have possessed, such mental capacities" (6). After reading his descriptions, in *Animal Liberation* and elsewhere, of the brutality humans can show to their fellow creatures, one might ask whether this approach would put vulnerable human and nonhuman animals at risk of callous and cruel treatment—and premature death. After all, does a newborn infant or senile adult qualify as a being "with a clear conception of the past and the (possible) future" (6)? What nonhuman animals have "these kinds of hopes and plans" (6)? In the end, might Singer's "mental capacity" standard blur differences of favor of a unifying category, the very gesture he deplores in speciesism?

In his discussions of the morality of killing, Singer adopts a number of unifying, reductionist, and value-laden labels. "Defective" is one of these. A word inherited, no doubt, from mechanistic science and philosophy, it is better applied to faulty vacuum cleaners than to human or nonhuman animal life. "Mental defectives" is even worse. Singer uses both, and often, for example: "killing a defective infant is not morally equivalent to killing a person. Very often it is not wrong at all" (*Practical*, 138), a statement that is not sufficiently critical, for one reason, because it does not attend to the multiple differences that prevail within the two groups he denominates: "persons" or "normal human adults" as one, "defective infants" as another. He also refers, without question, to "retarded humans" (even to "the grossly retarded 'human vegetable'"), apparently indifferent to the problematic conflation of disability with the man/nature hierarchy in the Western tradition, and of the long history during which animals and women were considered "defective" relative to "normal" adult male humans (see Singer *Practical*, 52–53; 75). In short, drawing a line between "normal" and "defective" is not as straightforward or "scientific" as Singer suggests.

Consider, for example, his discussion in chapter 4 of *Practical Ethics* of the morality of killing a Down's syndrome infant requiring surgery for a duodenal atresia and a septal heart defect, both of which can be quite readily corrected. Without noting that Down's syndrome individuals are not easily reduced to one prognostic type, and that many lead long, *happy*, and *pleasurable* lives, Singer contends that such an (Down's syndrome) infant "would never be able to live an independent life, or to think and talk as normal humans do" (73). Contrast this, he says, "with the casual way in which we take the lives of stray dogs, experimental monkeys and beef cattle" (73). Whether the casual killing of human "defectives" is likely to alleviate the casual killing of nonhuman animals is a question for critical animal studies. In later pages of this book, however, we will note that some contributors to animal studies are challenging at least four aspects of Singer's approach: one, relating to such terms as "defective," concerns the use of labels that group large numbers of living beings under a single concept that assimilates differences into sameness ("the animal" as chief among these); another is the theorizing of suffering as a power or capacity; a third concerns what for many is the Cartesian "mental capacities" standard on which suffering, in his ethics, is made to depend; and a fourth has to do with the "most like us" standard against which nonhuman animals are measured, this as resting on a conception of ethics as cost-benefit calculation, and of the "normal adult human" as its author and norm.

3

ANIMAL RIGHTS IN THE WILD

In *The Case for Animal Rights*, an extensive study first published in 1983, Tom Regan delineates what he calls his "rights view," its differences from other competing approaches including utilitarianism, and its application to animal welfare issues. The book provided philosophical grounding for the then-nascent animal rights movement, a social-political movement that, in his 2001 collection of essays, *Defending Animal Rights*, Regan describes as 10-million strong in the United States alone, and dedicated to "uncompromisingly abolitionist goals," seeking, for example, to bring an end to factory farming and to the use of nonhuman animals in biomedical research (40–41).[1] The abolitionist argument, strident at times, has marked the animal rights movement with controversy and violence, matters he addresses in the 2001 collection and in *Empty Cages: Facing the Challenge of Animal Rights*. For Regan, however, the abolitionist position cannot be qualified, which is one reason for his "anti-utilitarian" (43) stance. Whereas a utilitarian approach to animal welfare issues is "reformist," he says, for example in lobbying against gestation crates, battery cages, accelerated slaughter speeds, and the like, his animal rights position demands the complete cessation of human practices that routinely use nonhuman animals as sources of food or as models in scientific research (4). Hence the title *Empty Cages*, in which he argues that, "the truth of animal rights requires empty cages, not larger cages" (10).

And yet, in laying out this truth in *The Case for Animal Rights*, Regan argues that not all, but only some, nonhuman animals (and only some humans) count as "individuals," and thus as holders of fundamental rights. Does his line drawing itself compromise an abolitionist stance? And how

31

does his focus on "individuals" contend with situations where entire "wild animal" species and species habitats are at risk of extinction? With a view to these and related questions, this chapter outlines Regan's rights approach in the context of contemporary threats to the survival of wild animal species. The chapter title, "Animal rights in the wild," may be only provisionally useful, however. For although at one time, the word *wild* (OED: of animals or plants living or growing in the natural environment) was used in opposition to cultivated or human inhabited places, today, when such untouched "natural" places are not so easy to find, the meaning of "wild" has become less certain. Due largely to human encroachment, many spaces formerly considered "wild" now lie marginal to, or overlap, urban and industrial spaces. However, as Chris Philo and Chris Wilbert point out in their Introduction to *Animal Spaces, Beastly Places*, in numerous discourses, the human/animal dichotomy still prevails as a civilization/wilderness boundary that counts certain nonhuman animals as geographical "others," conceptually fixed in imagined "worldly places and spaces" different from those that humans occupy: companion animals or pets such as cats or dogs proximate to humans in settlement zones or cities; sheep, cows, and the like, in zones of agricultural activity; with other animals such as wolves and lions in more remote, "wilderness" lands (11). With these lines of division giving way to today's "complex entanglings of human–animal relations with space, place, location, environment and landscape" (4), Regan's individualist rights approach comes up against more "holistic" and ecologically based responses to issues of habitat depletion and species extinction.

Encroachment and Extinction

In her Introduction to *Hope for Animals and Their World*, Jane Goodall tells the story of growing up in Bournemouth, England with the dream of going to Africa to live with animals and write books about them, a dream that was realized when at age twenty-six, she was invited to visit her school friend in Kenya. Once in Kenya, she tracked down Louis Leakey, and through him was assigned the task of studying the behavior of chimpanzees in Tanzania's Gombe National Park. The chimpanzees of Gombe are still being studied, and Goodall would be there with them today, had she not attended a 1986 "Understanding Chimpanzees" conference, an event that she says changed the course of her life. The conference brought together for the first time field researchers from across Africa, and with them, the information that chimpanzee habitat was rapidly disappearing. "Right across their range, the chimpanzees' forests were being felled at a horrifying rate," Goodall writes; "they were being caught in poachers' snares, and the so-called bushmeat trade—the *commercial* hunting of wild animals for food—had begun. Chimpanzee numbers had plummeted since I began my study in 1960, from somewhere over a million to an estimated four to five hundred thousand (it is much less now)" (xx).

Having gone to the 1986 conference as a scientist, Goodall left as an advocate for chimpanzees and their vanishing forest home. Traveling around the world—since 1986, more than three hundred days a year—she has lectured, attended conferences, met with conservationists and legislators, and in the process, has come to realize that not just the forest homes of chimpanzees and other African animals are endangered, but "forests and animals everywhere. And not only forests, but all of the natural world" (xxi). Indeed, it is the case that biodiversity and species loss is occurring globally at an unprecedented rate. In *Ethics and Animals: An Introduction*, for example, Lori Gruen cites scientific estimates that by the year 2000, more than one hundred species of plants and animals were becoming extinct every day, a crisis that is intensifying, and that is caused by humans pursuing interests that threaten the survival of entire species and ecosystems (167). In *Hope for Animals and Their World*, Goodall brings actual stories to the statistics in her discussion of a number of mammal and bird species that were restored through captive breeding programs, either after becoming extinct in the wild (the North American black footed ferret, Australian mala wallaby, California condor, Chinese milu deer, North American red wolf, and Mongolian Przewalski's horse), or after being brought to the brink of extinction (golden lion tamarin of Brazil, American crocodile, peregrine falcon, North American burying beetle, crested ibis of China, North American whooping crane, Madagascar's ploughshare tortoise, Taiwan's landlocked salmon, and Canada's Vancouver Island marmot).[2] *Hope for Animals and Their World* includes additional stories of many animal species—from lynx to vultures, geese, and island birds—whose future is grave or uncertain.

What transformed Goodall from animal scientist to animal advocate was the arrival of the bushmeat trade that by 1986 was already devastating African ape populations. In *Eating Apes*—a book that Goodall suggests on the back cover, "you must read"—Dale Peterson examines the African bushmeat trade in graphic detail; his account, accompanied by Karl Ammann's revealing photographs, is at least as distressing as any graphic portrayal of factory farming. Peterson ventures at the opening of his book that there is nothing special about the fact that people living in and near forests of West and Central Africa happened to eat wild animal meat, "and probably have been doing so since human appetite began" (1). But with the loss of traditional ways in Africa, and with the arrival of modern weapons, modern population growth, modern cities, and with the unprecedented opening of African forests by European and Asian timber companies, something new has happened: "the consumption of wild animal meat has suddenly exploded in scope and impact, moving from what was until recently a subsistence activity to become an enormous commercial enterprise" (1). In the bushmeat trade, as in factory farming, huge industries, motivated solely by monetary profit, have swept aside traditional, environmentally conscious, methods of hunting and consuming forest meat in favor of a business of meat hunting and trading that abuses animals and eradicates their habitats. Peterson calls it a "wood-and-meat" business, in that commercialized logging, destructive of African forests, operates in tandem with the eradication of forest dwellers and the slaughter of apes for the meat trade (116).

Although situations vary from place to place, destruction of species, which goes hand-in-hand with the disappearance of "wild" spaces, owes inevitably to soaring human population growth, urban expansion, overfishing, deforestation, pollution, global warming, wanton animal slaughter for profit, and what Heather Beattie and Barbara Huck in their study of endangered species of Western North America *Wild West: Nature Living on the Edge*, refer to as modernity's still prevalent attitude that: "Nature exists for the use and pleasure of humans" (15).

Ethical Holism?

By her own account, Jane Goodall's increasing awareness of the global scope of animal habitat disappearance only confirmed her love for, and commitment to, not just chimpanzees, but all animals, all of the natural world. "There is no sharp line dividing us from the chimpanzees and the other apes, and the differences that exist are of degree, not of kind," she maintains in *Hope for Animals and Their World* (xx). By the same token, as evidenced by her efforts over the past many years to prevent destruction of the natural world, Goodall draws no sharp line between humans and life forms other than chimpanzees. In dedicating herself to protecting species from extinction, she works on behalf of insects as well as birds, bears, amphibians, reptiles—and trees, all species of plant and animal life that are threatened by "[h]uman population growth, unsustainable life-styles, desperate poverty, shrinking water supplies, corporate greed, global climate change" (xxiii), all these and more. Hers is as good an expression as any of what has been called an ecological or "holistic" approach to ethics, an approach that, even if it does not fare well with Peter Singer or Tom Regan, has great appeal for environmentalists and wildlife resource planners. As geographer William S. Lynn frames this approach, since the differences that exist between human and nonhuman animals "remain distinctions, not dichotomies," and since humans are "related to yet different from the other species with whom we inhabit the earth," ethics is inescapably "geoethics," an ethics that considers humans as embodied and "ennatured, that is, situated in nature's rich web of life-forms and life-forming processes" (283).

For wildlife conservationists and managers, perhaps the most influential example of such holistic thinking is Aldo Leopold. As a thinker, teacher, activist, and writer of such texts as the famous *A Sand County Almanac* (1949), Leopold both defined and modeled wildlife resource management for future decades of the twentieth century, always out of his profoundly ecological view of the inclusiveness of humans in nature. As Curt Meine notes in "Moving Mountains," Leopold also affirmed the interdependence of science, ethics, literature, and art, thus the importance of interdisciplinary approaches to animal studies.[3] Like Goodall, Leopold was a boundary-crosser who, in favor of the "biotic" whole, disregarded ontological and disciplinary lines of division. Yet, precisely because it negates line drawing, his is a position that, according to Tom Regan, the rights view "cannot abide." In *The Case for Animal Rights*, referring to Leopold's ethics as "environmental fascism," Regan states that Leopold's view and the rights view "are like oil and water: they don't mix" (362). This is a point to which we will return shortly.

Alexandra Morton, like Leopold and Goodall, would likely belong to Regan's "fascist" group, for she, too, eschews line drawing in favor of an ecological view. A marine biologist, Morton arrived on the central Pacific coast of Canada in 1984 to study the acoustics of orca whales and the return of the Pacific white-sided dolphin to the region after an absence of some seventy years. In the late 1980s, however, after industrial salmon farming was introduced to the area, she began to notice its deleterious effects on the region's marine ecology, subsequently enlarging her research focus to include documentation of the impact of factory fish farming on various ocean species and engaging in activist efforts to preserve marine ecology. When her activist efforts failed to prompt politicians to monitor or address the situation, Morton put on her scientist cap and undertook extensive research and publication on the damaging side effects of industrial fish farming, posting her research results on the website of the Raincoast Research Society that she founded: a substantial body of evidence to support the case that the factory farming of fish poses significant environmental risk and has already resulted in an alarming decline of ocean fisheries and ecosystems—a case that has now been made and well documented by others (see for example Stier and Hopkins). In the small coastal community of Echo Bay, accessible only by floatplane, Morton and some forty other people live in floating houses, raising their children as part of the ocean ecosystem, rather than as spectators removed from it, and educating children and others in the ethics of species co-existence (see Alexandra Morton's Raincoast Research Society homepage at www.raincoastresearch.org/home.htm).

Farther down the North American west coast where, as a consequence of the invasion by people and cities of California wildlands that had once been cougar habitat, the outlook for cougars in the region is bleak—unless, as Andrea Gullo, Unna Lassiter, and Jennifer Wolch argue in "The Cougar's Tale," humans and cougars can learn how to co-exist. "The plight of animals worldwide has never been more serious than it is today," Wolch and Jody Emel write in their Preface to *Animal Geographies: Place, Politics, and Identity in the Nature-Culture Borderlands.* This reality is largely obscured, they suggest, by the progressive elimination of animals from everyday human experience, "and by the creation of a thin veneer of civility surrounding human-animal relations, embodied largely by language tricks, isolation of death camps, and food preparation routines that artfully disguise the true origins of flesh food" (xi). In the interests of bringing animals back into everyday human experience, and in the process removing the threat of extinction, Wolch, professor of geography and urban planning and director of the Center for Sustainable Cities at the University of Southern California, directs her efforts to formulating a trans-species urban theory and praxis.

Wolch coins the term *zoöpolis* to refer to the kind of urban center that would be designed specifically for coexistence of humans and nonhumans, including "wild," animal dwellers. In an essay by that title, she points out that over the past many decades, theories and practices of urbanization have contributed to disastrous ecological results: proceeding with disregard for nonhuman life, urban development has taken over so-called "empty" wildlands, threatened entire ecosystems and species, denaturalized the environment, impoverished soil, and in general, worked in the interests of monetary profit and of humans alone, at the expense of wild, feral, and captive animals ("Zoöpolis," 119–120). Arguing for a nondiscriminatory ethics and practice "of caring for animals and nature," she envisions the zoöpolis model: a reintegrated, nonspeciesist, city that might serve to curb "the contradictory and colonizing environmental politics of the West as practiced both in the West itself and as inflicted on other parts of the world" (124). The zoöpolis is nonutopian, she insists, citing several cities in the United States that are moving in the direction of a trans-species urban practice, even if this coexistence movement has as yet been poorly documented and undertheorized (131).

Animal Rights

In *Empty Cages*, as he has done elsewhere, Tom Regan presents his rights view in an autobiographical narrative: growing up as a meat-eater who loved fishing, dissected animals in biology labs, and even worked as a butcher during his college years, he did not consider the idea that killing and eating animals might be morally wrong until, as a beginning philosophy teacher opposed to the Vietnam War, he encountered the pacifist writings of Mahatma Ghandi. Gradually, after exploring the philosophical and political legacy of human rights, he arrived at the conviction that animals possess rights, too. Coming to this position was not, for Regan, a matter of challenging the modern concept of rights per se. Indeed, in line with the liberal tradition, he defines a right as a "possession" that grants negative freedom to its holder, something like a "No Trespassing sign" (38). A right is like a fence that walls off an individual and his or her body from interference by others. Rights, he says, are adversarial: when people exceed their rights by violating ours, "we act within our rights if we fight back, even if this does some serious harm to the aggressor" (39). Moral rights "breathe equality" (39). They are like "trump" cards: they have greater weight than other important values (40). They are claims that an individual is "owed or due" (41) and thus they are not pleas but "demands," as he puts it, "for justice" (41). All of these components of rights theory, in line with the liberal individualist tradition, sit well with Regan. The only question for him is that of "the individual" to whom equality is due. Are nonhuman animals also individuals in possession of rights? His short answer to the question is: yes, if they are "just like us in being subjects-of-a-life" (53).[4]

Regan presents "subjects-of-a-life" as a nonspeciesist term that provides for the inclusion of nonhuman animals as "individuals" possessing fundamental rights: whether nonhuman animals have rights depends on whether they qualify as "subjects-of-a-life." In the updated Preface to his seminal study, *The Case for Animal Rights*, he acknowledges that the term necessitates line drawing, because not all animals are "subjects-of-a-life" (xvi). Rather, the term applies only to "mentally normal mammals of a year or more" (xvi). Later in the book, he explains that the term restricts his usages of the words "human" and "animal":

> Henceforth, when either humans or animals are referred to, it is to be assumed that those to whom we refer are individuals *well beyond* the point where anyone could reasonably "draw the line" separating those who have the mental abilities in question from those who lack them. Unless otherwise indicated, that is, the word *humans* will be used to refer to all those *Homo sapiens* aged one year or more, who are not very profoundly mentally retarded or otherwise quite markedly mentally impoverished (e.g., permanently comatose). And unless indicated otherwise, the word *animal* will be used to refer to mentally normal mammals of a year or more. (78)

Extending animal rights to mammals, because, as he writes in *Empty Cages*, "other mammals are like us" (59), would include thousands of nonhuman species, along with some ten thousand species of birds— *Empty Cages* also includes birds as subjects-of-a-life, in that birds are "like us" in morally relevant respects (60). But Regan's term, subjects-of-a-life, does not apply to species. In keeping with his definition, rights are the possessions of *individuals*—individual human and nonhuman mammals and birds who are at least one year of age and "mentally normal." A number of questions arise here. For example, how, in every case, would (apparently empirical) assessments of "mentally normal" be made, and by whom? "Perception, memory, desire, belief, self-consciousness, intention, a sense of the future—these are among the leading attributes of the mental life of normal mammalian animals aged one year or more," Regan explains in *The Case for Animal Rights* (81). Do these traits apply equally well to a hedgehog as to a human being? Some critics of Regan's approach think not, citing his human-centered "most like us" standard as in the end an anthropocentric and even speciesist argument that, with its focus on the individual and on "mental life," perpetuates Cartesian self/other, mind/body hierarchies.[5] And despite his distinction between "moral agents" and "moral patients," Regan's dual requirements—both "mentally normal" and "one year of age or more"—eliminate human infants, along with many disabled and elderly human individuals, those at either end of the life spectrum who are most vulnerable. With respect to such others as these, Regan does make some concessions, albeit paternalistically: for example, in *The Case for Animal Rights*, he refrains from endorsing abortion and infanticide by advocating that infants and viable human fetuses be given "the benefit of the doubt, viewing them *as if* they are subjects-of-a-life, *as if* they have basic moral rights, even while conceding that, in viewing them in these ways, we may be giving them more than is their due" (320). Perhaps it is such both-and-wavering—at once withholding subject-of-a-life status and conceding "the benefit of the doubt"—that proves most problematic where Regan's rights view contends with ethical issues involving wild animals.[6]

The Rights of Wild Animals

Regan is categorical in distinguishing his approach to animal ethics from that of Peter Singer. What is right or wrong, he insists, depends not on the consequences of given actions. Moreover, utilitarianism "seems to imply that good ends justify whatever means are necessary to achieve them, including means that are flagrantly unjust" (2001, 16). His rights view, in opposition to this, argues that, "the principle of respect for an individual's rights should not be compromised in the name of achieving some greater good for others" (16). In short, for Regan, Singer's understanding of animal liberation is "profoundly mistaken" (36). Curiously, nonetheless, both Singer and Regan resort to an anthropocentric "like us" standard in determining which nonhuman animals count as having equal interests or rights.[7] Nowhere is the standard more sharply focused than in arguments for granting rights to great apes, a context in which, oddly enough, Singer too adopts the language of rights.

Along with Paola Cavalieri, Singer is editor of *The Great Ape Project: Equality Beyond Humanity*, a collection of essays by contributors, including Regan, who are committed to the inclusion of nonhuman great apes within the moral community. Singer and Cavalieri explain in their Preface that the Great Ape Project seeks to extend basic human rights to "the species that are our closest relatives and that most resemble us in their capacities and in their way of living" (1). Sufficient evidence now exists that chimpanzees, gorillas, and orangutans are "intelligent beings," they maintain, and the time has therefore come for reassessment of their moral status and acceptance of them as "persons" (1–2). The collection opens with a statement titled "A Declaration on Great Apes," which calls for the extension to great apes of certain "human" rights, including the right to life, protection of individual liberty, and prohibition of torture. The Declaration states that great apes would be the first nonhuman animals to be regarded as "members of the community of equals," and rightly so, for they are "the closest relatives of our species," and they also "have mental capacities and an emotional life sufficient to justify inclusion within the community of equals" (5).

Similarly, in his contribution, Regan takes as his point of departure increasing evidence attesting to "the intellectual abilities" of great apes, and the possibility that these primates have the ability to understand and use language. After arguing that we are morally obliged to protect the "individual interests" of both human and nonhuman "incompetents" (great apes, for example, who, although they have known preferences, are unable to give or withhold informed consent), he turns to a reiteration of his position that those who possess inherent value are subjects-of-a-life, a standard that necessitates the line drawing but that includes individual mammals such as great apes. They qualify as subjects-of-a-life, so great apes are morally equal to humans, and should not be used for harmful research or experimentation, any more than should humans (194–205). Here as elsewhere, Regan's "like us" standard puts the onus on similarity "defined in human terms," and "theorized via an overly cognitive and rather disembodied ontology" (Twine 28; 29).

The "like us" standard espoused by both Singer and Regan has been adopted widely in animal studies. For example, Dale Peterson, whose book *Eating Apes* is considered above, refers to apes as our "sibling species," who "observe the world through eyes and faces like ours, who manipulate the world with hand and bodies like ours, who display emotions entirely recognizable to us" (1). Ape brains closely resemble the brains of humans, so they share something of the human mental world (7; 15), and what for him is of great importance, apes are "closer to us" than we think in their ability to communicate through language. On this point, Peterson refers to recent attempts, notably that of Allen and Beatrix Gardner at the University of Nevada in Reno, to teach American Sign Language to apes—to one chimpanzee they reared from infancy "within some approximation of a human household," and to four others with whom their graduate student Roger Fouts works at an enclosure at Central Washington University. All have progressed "well enough to have conversations with human deaf signers" (15), which confirms for Peterson that "apes share something of our human mental world" (15).[8] The argument is not far removed from that made by Descartes in the seventeenth century, when he declared language use ("real speech") to be evidence of the presence of thought in a body. Brute animals lack "the perfection of using real speech," he writes in one of his letters, "that is to say, of indicating by word or sign something relating to thought alone and not to natural impulse. Such speech is the only certain sign of thought hidden in a body" ("Letter to More" 366).

It is as a conservationist dedicated to saving the world's remaining apes, rather than as a rights theorist, that Peterson offers his "like us" argument. Similarly, while Goodall is a signatory to the Declaration on Great Apes, she is neither a rights theorist nor an advocate of the "like us" argument that establishes the intellectual ability of "normal" adult humans as the standard against which the moral worth of other species should be measured. Rather, hers is a plea for conservation and inclusiveness, made on behalf of apes, and of all the world's plant and animal life. This is a point on which Goodall, and environmentalists in general, part with Regan (as well as Singer), who allocate rights (or interests) only to *some* human and nonhuman animals. As Regan points out in *The Case for Animal Rights*, his rights view "is about the moral rights of individuals," and it "does not recognize the moral rights of species to anything, including survival" (359). He goes on to say this: "That an individual animal is among the last remaining members of a species confers no further right on that animal, and its right not to be harmed must be weighed equitably with the rights of any others who have this right" (359). Thus, if the rights view supports efforts to save endangered species—and Regan claims it does—the reason for this support is based on the rights of those individual animals who qualify as subjects–of–a–life, whether they are members of an endangered species or not.

It follows, as mentioned earlier in this chapter, that Regan's rights theory has little in common with the views of Aldo Leopold, which suggest that individual interests may be sacrificed for the greater good of the biotic community. According to Regan, Leopold's "environmental fascism" (*Case*, 362) is wrong in suggesting that individual rights can be outweighed by aggregative considerations, including considerations of what will or will not contribute to the integrity of the biotic whole. And since, as Regan insists, "the rights view categorically denies that inanimate objects can have rights" (362), there seems little prospect for a land ethics derived from his theory. Rights-holders are individuals, he insists, not "inanimate" natural objects. Moreover, he ventures that if respect were shown to the rights of individuals, the biotic community would also be preserved. "And is not that what the more holistic, systems-minded environmentalists want?" (363; see also *Defending*, 19–22). About wildlife resource management, Regan has nothing good to say, not surprisingly, because the field is given to regulation or "management" of such activities as hunting, fishing, and trapping, rather than to abolition of them (see *Case*, 354–355).[9] Instead of recognizing the rights of individual wild animal subjects-of-a-life, wildlife management policies, in his view, are basically utilitarian. "No approach to wildlife can be morally acceptable if it assumes that policy decisions should be made on the basis of aggregating harms and benefits. In particular, these decisions should not be made by appeal to the minimum harm principle" (356).[10] On all counts, the rights view is at odds with "reform" or "regulation" of any sort. "What it recognizes is the prima facie right of individuals not to be harmed, and thus the prima facie right of individuals not to be killed" (359).

One can glimpse some of the political and social implications of Regan's view, simply by reflecting on his abolitionist list, which includes the following: empty cages; halt animal farming, including the factory farming industry; end all animal experimentation; abolish trapping, sport, and recreational hunting, including fox hunting; prohibit predator control programs; outlaw commercial whaling and the slaughter of seals; and declare vegetarianism to be obligatory. Consideration of the legal dimensions of extending human rights, even to some nonhuman animals, has already generated an extensive literature that includes such texts as *Animal Rights: Current Debates and New Directions*, edited by Cass. R. Sunstein and Martha C. Nussbaum, in which ten of sixteen essays come from lawyers, or from professors or philosophers of law; American lawyer and animal rights proponent Gary Francione's *Animals, Property, and the*

Law; and Joan E. Schaffner's *An Introduction to Animals and the Law*. Curiously enough, such texts as these acquaint readers with various regulatory developments (animal welfare acts, anti-cruelty laws, regulations for use of animals in research), the sorts of reform initiatives that Regan's rights view opposes.

His uncompromising stance on animal rights is unlikely to gain wide acceptance. After all, in any sustainable social or ecological system, not every human or nonhuman "subject-of-a-life" can lay absolute claim to individual rights. Present day courts are already overwhelmed by the litigation of clashing rights claims. Moreover, the adversarial nature of rights, their perpetuation of a liberal humanist understanding of the "person" or subject, and the "like us" standard on which Regan's subject-of-a-life concept rests, have all come under recent critique from within animal studies. Finally, since Regan's rights theory is itself the product of modern Western individualist culture, and the framework that facilitated that culture's colonizing enterprise, it is unlikely to succeed as a cross-cultural, any more than as a cross-species, ethics based on respect of differences.

4

ANIMAL EXPERIMENTATION

In the Lab and on the Farm

In 1977, British art historian, critic, and novelist John Berger wrote one of his most compelling essays, "Why Look At Animals?" (published in his 1980 book *About Looking*, also available in the collection of his essays that was published in 2009 in the Penguin *Great Ideas* series under the title *Why Look At Animals?*). Berger opens his essay this way: "The 19th century, in western Europe and North America, saw the beginning of a process, today being completed by 20th century corporate capitalism, by which every tradition which has previously mediated between man and nature was broken" (1). As a result of this rupture, Berger goes on, "animals have gradually disappeared. Today we live without them" (9). That is, the animals that at one time "constituted the first circle of what surrounded man" (1) no longer exist. They have been eradicated, whole species at a time, poisoned by toxic waste and pollutants, killed by the millions every year in research facilities or testing labs, processed as raw material, locked up in zoo and circus cages, turned into chimeras, experimentally altered and engineered as farm factory commodities. This is why animal historian Jonathan Burt suggests that "[t]he history of animals is, among other things, the history of the disappearance of the animal" (290). If we are not living *after* "the death of animals," we are fast approaching that end, Berger and Burt suggest, and as both point out, this is the result not only of human practices that have led to the acceleration of species extinction and global loss of biodiversity, but also to attitudes toward animals that enable these practices. Changed ways of thinking about, or as Berger puts it, of "looking at," animals are behind their death and disappearance today. The present chapter considers this situation with respect to the history, current status, and ethics of animal experimentation.

Historical Legacies

Insofar as "experimentation" is defined as the use of nonhuman animals to gain knowledge of human anatomy and physiology, the practice has a long history, reaching back at least to ancient Greece and including not only the dissection of dead animal bodies (which likely has a longer history still), but also animal vivisection (from the Latin, *vivus* = alive, living + *sectio* = a cutting, vivisection refers to the action of cutting open a living body, not for surgical purposes but in the interests of improving knowledge or of performing scientific research). For example, the works of Aristotle (384–322 BCE), almost a quarter of which comprise treatises on biology (including *History of Animals, Parts of Animals, Movement of Animals*, and *Generation of Animals*), indicate the importance that animal experimentation already had by this time. Aristotle's detailed descriptions of vascular, neurological, digestive, respiratory, reproductive, muscular, and skeletal systems are based on his animal dissections. He never dissected a human body, for as historian of science Anita Guerrini notes in *Experimenting with Humans and Animals: From Galen to Animal Rights*, such a dissection would not have been allowed in the Greece of his day (10). However, he did dissect numerous dead animals, sometimes killing animals for that purpose, and in addition, "he experimented on (but did not cut open) live animals, arguing that living function could be explored only in the living" (10). No doubt there was cruelty involved in his research, for instance in his starving and then strangling the animals he used for vascular dissection.[1] In Aristotle's view, however, he was engaged in a laudatory pursuit: the quest to know for what purpose everything in nature has been created as it is. He approached each of the "parts of animals" as purposeful and as according with an overall, divinely ordained, *telos* or end. And although his top-down "chain of being" cosmology located humans above animals (and men above women) in the hierarchical order of things, he still assumed that an analogy exists between human and animal bodies (only humans possess intelligence, he said, which raises them above animals).[2]

Guerrini notes that two of Aristotle's followers, the physician Herophilus (ca. 330–ca. 260 BCE) and his younger colleague Erasistratus (304–245 BCE), practicing in Alexandria, were given permission by their king to cut into a living man, a condemned criminal sentenced to death, not to cure him but in order to look inside (6). The works of these physicians are lost, but we know from the second century CE Roman historian Celsus that they also dissected human cadavers and practiced animal dissection and vivisection. Celsus elaborated arguments for and against vivisection, and in an early version of "consequentialist" ethics, he "offered the argument that the sacrifice of a few for the many was justified. Celsus lived and wrote in ancient Rome, where a criminal could 'pay' for his crime by making his body useful to the community" (8). If this argument justified human vivisection, it no doubt condoned similar use of "lesser," nonhuman, animals as well.

But, following Aristotle, it was the Roman physician Galen (ca. 130–210 CE), deriving his ideas on human anatomy and physiology solely from animal dissections and vivisections, whose work dominated Western medical theory and practice for centuries. With Galen, animal dissections and vivisections became public events, "the medical equivalent of a gladiatorial combat," in Guerrini's words (13).[3] Yet Galen never dissected a human cadaver. Rather, he based his work on the analogy Aristotle proposed between animal and human bodies. Relying on this, "the best he could do was to dissect a Barbary ape, perhaps with a naked slave nearby to serve as a comparison" (14). As long as the analogy prevailed, animal experimentation (dissection and vivisection) was routinely used as the method for obtaining knowledge about the structure and function of human bodies. Not until 1543 and the publication of Andreas Vesalius' *Fabrica* (*De humani corporis fabrica libri septem*, that is, *Seven books on the structure* ["*fabric*," or "*workings*"] *of the human body*), was the Aristotelian analogy, along with the authority of Galen, overturned. It was a pivotal moment—in the same year, Copernicus published his *De revolutionibus*— but it did not spare animals from the scalpel. For one thing, as the last chapter of the *Fabrica* makes clear, Vesalius considered vivisection (his performed on apes and dogs) to be an essential instructional method for demonstrating physiological processes, such as the movement of the diaphragm in breathing. He used illustration by vivisection in part to point out and correct certain of Galen's errors, in part to advance the study of comparative anatomy. And he raised no questions about the suffering animals endured in vivisection: their purpose was to serve as anatomical teaching and research devices. Animals had satisfied this purpose before, but in the era of Vesalius, attitudes toward animals began radically to change.

As Europe's leading early-modern anatomist, a reputation solidified by his publication of the *Fabrica*, Vesalius introduced the spectacle of public human, rather than animal, dissections.[4] Thus the frontispiece of the *Fabrica*, showing Vesalius himself dissecting the body of a human female (this woman was taken from the gallows), actually cutting open her uterus, represents an anatomical first, for as Vesalius writes in the book, "Galen never dissected a female uterus" (142).[5] The anatomy takes place before a crowd of spectators in what appears to be a theater, the dissection very much a performance with the anatomist, Vesalius, occupying center stage, shown with his hand on the corpse and *speaking for himself*, rather than reading from an authoritative work by Aristotle or Galen. As is suggested by removal of the rhesus monkey and dog in the *Fabrica* to the sides and below the dissection table, beginning with Vesalius, animals are no longer equal to humans as sources of anatomical knowledge. Dismissing the long-standing principle of analogy, Vesalius maintains that Galen, was "deceived by his monkeys," and that he failed to note "the many and infinite differences" that pertain between humans and animals (321). Here, one hundred years before Descartes introduced his mind/matter binary, Vesalius posits an "infinite" human/animal difference, displaces the authority of tradition, and dares to speak for himself. With the Aristotelian "chain of being" beginning to collapse, the new autonomy Vesalius claims for the speaking self (the emerging subject) soon displaces not only the transcendence of tradition, but also of tradition's deity. In the Vesalian anatomy theater, we glimpse an early inauguration of the subject-centered discourse that has now come under question in critical animal studies.

Still, the modern tradition of experimentation really begins with Descartes (1596–1650). Born in France and educated by Jesuits in what he describes in his *Discourse On Method* as the best schools Europe had to offer, he determined nonetheless, as did Vesalius (whose *Fabrica* Descartes studied), to throw off the authority of tradition in favor of speaking for himself, accepting as true only what he learned firsthand. Significantly, when Descartes relocated from France to the Netherlands in 1629, his move coincided with the transfer of the center of anatomy from Padua to Leiden and then Amsterdam, the city in which he settled. There, he turned his attention to anatomy, regularly visiting the butcher stalls in the Kalverstraat quarter in which he lived and hauling animal cadavers back to his flat for dissection, learning anatomy by his own hand—and urging his readers, similarly, to learn for themselves. "I would like anyone unversed in anatomy to take the trouble, before reading this, to cut up the heart of some large animal," he writes in the *Discourse On Method* (134, translation altered). Descartes also practiced vivisection, as we know from his *Description of the Human Body and of All Its Functions*, where he gives account of his vivisection of a dog (317). As we note in earlier pages, however, Descartes' major legacy to modern experimentation consists not in his awkward and error-ridden anatomical and physiological accounts, which were soon enough surpassed, but in his mind/matter, man/animal dualism. With his assertion that animals, all animals, are only machines—as much without the capacity for real speech as for real pain—he provided a wholesale justification for the use of animals, living or dead, in experimentation and research.

Descartes' contemporary, the English physician William Harvey (1578–1657), who studied medicine at Vesalius' alma mater, the University of Padua, set out to prove that the heart is the central organ of the body, not the liver as Galen had argued. Like Descartes, Harvey approached the heart, in its structure and function, as a mechanical entity, and so great was his enthusiasm for advancing the principles of mechanistic science, that the brutality involved in his cutting open and experimenting on scores of live animals did not give him pause. "Animals, to Harvey, served a purely instrumental function," Guerrini writes, "and if the question of cruelty occurred to him, he never expressed it" (31). To establish the mechanical nature, and direction, of blood circulation, Harvey observed dogs in vivisection: indeed, during the seventeenth century, so many physiologists vivisected and otherwise experimented on dogs that, as historian of science Richard Westfall remarks, "one is sometimes surprised that the canine species managed to survive" (88). In addition to dogs, Harvey dissected, vivisected, or experimented on birds, rabbits, cats, and in his studies of embryology, on numerous deer. His work contributed immensely to the seventeenth-century application of mechanics to anatomy and physiology, and eventually to solidifying the view of the body as but a machine.

Many other seventeenth-century scientists participated in advancing this view—always, it seems, at the expense of animals. In London, England, for example, a group of scientists, among them Robert Boyle (1627–1691) and Robert Hooke (1635–1703), formed the Royal

Society, advancing the new mechanical science through use of numerous living and dead animals. Guerrini provides a description of some of their experiments: including ligature of blood vessels of animals in vivisection, or injection of various substances to demonstrate structure; injection of poisons and paralytic agents to show their mode of action; use of the vacuum pump invented by Boyle and Hooke to asphyxiate animals and thus demonstrate that they need air to survive; cutting open the thorax of animals to determine the mechanisms of respiration; and experiments with animal-to-animal blood transfusions, dog-to-dog or cross-species, with sheep and calves as well as dogs (37–41).

Already in the seventeenth century, then, with rapid growth of the new science, such reduction of animals to experimental objects at the disposal of experimenting subjects had come to mark the beginning of their end, or of what John Berger calls their "disappearance." For many, the idea of an animal machine held out the promise that science could achieve full knowledge of the mechanism's workings, and with that, a vast increase in human control of, and power over, nature. Knowledge, once valued for its own sake, was now valued for its usefulness in augmenting human power, such that, in the work of seventeenth-century philosopher Francis Bacon (1561–1626), for example, mastery through positive science became the enlightened man's ideal. In Bacon's words, spoken to his servant and quoted by Richard Westfall in *The Construction of Modern Science*, "The world was made for man, Hunt, not man for the world" (118). Westfall notes that Bacon summarized this view in the expression, "the Kingdom of Man," which referred to "the physical world, the domain intended for man by God, an inheritance into which he can enter only by the path of natural science. To Bacon, knowledge was power, power by man can subject nature to his will and force her to serve his ends" (188).

The Cartesian model of the "animal machine" carried well into the nineteenth and twentieth centuries, and with it, the idea that, without moral qualms, animals could be used for scientific research. In the nineteenth century, British naturalist Charles Darwin's evolutionary theory might well have upset the dualistic model proposed by René Descartes, for Darwin (1809–1882) suggested not a dichotomy, but a commonality, between humans and animals: a long evolutionary process through which (by means of subtle differences, not a dramatic rupture) one emerged from, and remained related to, the other. Yet Darwin's *On the Origin of Species By Means of Natural Selection* (1859) also suggested that what he called *natural selection* involves nature's preservation of those subtle differences that, while randomly introduced, turned out to have survival value. In his *Descent of Man*, he took the point further, proposing that selection could also take place through differences introduced by man. Why not breed favorable traits—in animals, and indeed, in humans, thereby giving man greater power and control over the evolutionary process, making the process useful to human and scientific ends? Focusing on this question, Darwin's successors increasingly adopted his work in support of the philosophical and scientific paradigm that elevates man, as controller, over the natural and animal world, even over the processes of natural selection.

We will not delve into the eugenics movement that developed out of Darwin's theory, his cousin Francis Galton an early proponent of breeding for "genius" and, overall, for the characteristics favored by the nobility of his day. The movement was strong from the late nineteenth century into the 1930s, when revelations about the Nazi eugenic program brought it to a halt. Revived in the 1950s after the discovery of the structure of the gene by James Watson and Francis Crick, the eugenics movement reconstituted itself as a "genetic counseling" enterprise, working in the beginning to assist family planning through crude sketching of "family trees"; radically transforming its family planning in the 1970s and 1980s with the advent of such technologies as amniocentesis, ultrasonography, *in vitro* fertilization, and recombinant DNA; and most recently, undertaking (primarily on animals) a multitude of experiments, such as cloning, trans-species transplantations, and manipulations of early developing embryos. The ongoing development of molecular and genetic techniques continues to change the face of animal experimentation, a point discussed further in the next section of this chapter.

In the meantime, it is safe to say that institutionalized experimentation on animals is now entrenched around the world. Widely regarded as basic to the biological sciences, medicine, and experimental psychology, it involves everything from pharmaceutical, cosmetic, and toxicology testing, head injury research, xenotransplantation, and behavioral and radiation outer space experiments. The animal studies literature is replete with references to such research, all of which "does involve, at least some of the time, the torturing of animals" as Robert Garner states it succinctly (121). Here, but two examples might be mentioned. The first was brought forward in Canada's national newspaper, the *Globe and Mail*, in a Sunday, March 12, 2010, feature by Martin Mittelstaedt titled "Eureka!

Less really is more – deadly" (Section F1–7). The feature reports on recent testing of the long-held theory that, as exposure to something harmful increases, so do the harmful effects. Over the past twenty years, Mittelstaedt explains, scientists have discredited this theory, for instance, by experimenting on mice, feeding them the chemical bisphenol A (BPA), an ingredient found in some plastic bottles and children's plastic toys. One University of Missouri scientist fed pregnant mice less than one-tenth the BPA dose that Health Canada considers safe, only to discover that, although the resulting male offspring looked normal at birth, they entered adulthood with grossly enlarged prostate glands. With hundreds of thousands of such studies of chemical contaminants recorded, vast numbers of laboratory animals, a majority of them rodents, have been used to relate doses of harmful chemicals to degrees of toxicity.

In a second example, the toxicity in question concerns exposure to radiation rather than to chemicals, and monkeys replace rodents as the preferred animals on which to experiment.[6] The United States of America's National Aeronautics and Space Administration (NASA) recently announced its intention to fund an experiment in which as many as thirty squirrel monkeys will be exposed to harmful doses of ionizing radiation. Opposing the project, neurologist and public health specialist Aysha Akhtar writes in the *Huffington Post* ("NASA's Wrong Stuff") that, while astronauts on extended trips through deep space are continuously exposed to low levels of radiation, the NASA experiment will expose the monkeys to a single, large dose of radiation over a period of just a few minutes. NASA's stated reason for doing this is to observe how the radiation exposure affects the monkeys, and thus, how it might similarly affect astronauts' neurochemistry, cognition, and behavior. One point that Akhtar raises in protesting this experiment is that it will not provide much knowledge on the effects of space radiation on astronauts, who are not exposed to the single, high-level dose that NASA proposes to use on the monkeys. Akhtar argues moreover, that "four decades of radiation experiments on monkeys have cost thousands of animals their lives and taxpayers millions of dollars, yet have provided hardly any useful information about the effects of space radiation on humans." More than poor results matter to Akhtar, however, who expresses grave concern over the suffering the experimental monkeys will endure, not only as a result of being irradiated, but also from their confinement and manipulation in the lab, in an unnatural environment, deprived of social interaction, manipulated, physically tortured—five days a week, and for a period that NASA says will last at least four years.

Ethical Responses to Animal Experimentation

How have ethicists responded to such experiments as these? Answering just this question in the *Encyclopedia of Bioethics*, Peter Singer suggests that opponents of animal experimentation divide into two groups, reformers and abolitionists, the two groups we have already identified with Singer himself and Tom Regan. For the abolitionists, Singer suggests, the end does not justify the means; this position, which Regan espouses, would oppose even a single experiment that, say, held out the prospects of finding a cure for cancer. For utilitarian reformers, on the other hand, and these include Singer, some animal experimentation may be justifiable, even if most is not. Reformers claim (and here Singer refers to himself) that alternative methods, not involving animals, could replace much of the animal experimentation now being done. On the moral status of animals, he states his anti-speciesist argument that infants and "mentally retarded humans" actually fall below some adult dogs, cats, pigs, or chimpanzees "on any test of intelligence, awareness, self-consciousness, moral personality, capacity to communicate, or any other capacity that might be thought to mark humans as superior to other animals" (82). He suggests then that supporters of animal experimentation should be prepared to explain why experimentation on "mentally retarded humans" is not equally justified (82). In other words, in this early text as in his later works, Singer frames the ethics of animal experimentation along the lines of *either* utilitarianism *or* animal rights, and he invokes the "capacities" argument that both he and Regan continue to advance. The argument is hierarchical and anthropocentric in positing intelligence, self-consciousness, and other "like us" features ("us" referring to mentally "normal" adult humans), as necessary to moral worth, and by way of countering speciesism, the argument suggests that humans with lesser capacities have less moral worth than "us," and might then replace some animals as candidates for research.

Insofar as animal experimentation remains fenced in to the either/or ethical options Singer outlines in this early piece, either his utilitarianism or a variant of Regan's rights view, it demonstrates an impasse beyond

which animal studies has yet to move. For one thing, this either/or framing contains animal ethics to philosophy, and more specifically, to a particular specialization within analytic philosophy—implying that other
disciplines and other philosophical traditions have nothing significant
to contribute where the thinking of animal ethics is concerned. Indeed,
as will emerge in subsequent chapters, this is just what Singer argues—a
stance that is reconstituted in numerous anthologies and single-authored
animal studies texts that largely contain theories of animal ethics to
either utilitarianism or rights. This containment has only recently been
challenged on several fronts, and for the reason that it impedes the development of more *critical* animal studies, one that does not rely, without
question, on the tenets of a liberal humanist framework. Some of these
challenges are important for the questions they pose and the alternatives
they present to animal studies. Accordingly, in the following section of
this chapter and the following chapters of this book, we will draw on a
number of works that attempt to make animal studies a more open, and
openly self-critical, field.

Factory Farming Is Animal Experimentation

To read sociologist Richard Twine's, *Animals as Biotechnology: Ethics,
Sustainability, and Critical Animal Studies,* is to realize, with John Berger,
that, at least in agriculture, animals have long since disappeared. Peter
Singer and Jim Mason use the term biomachines with reference to
factory-farmed animals; Twine, who concentrates his study on genetic
and molecular techniques, contends that these animals have themselves
become "biotechnologies." Discussions of animal experimentation typically focus on the medical or scientific research lab, but Twine's book
takes factory farming itself to be a locus, perhaps even the central locus,
of today's animal experimentation. The book offers an invaluable introduction to four particular sets of techniques that are emerging in the
global factory-farming industry as means for engineering and "editing"
new sorts of animals: marker-assisted selection (identifying loci or markers within a specific area of an animal's genome); genomic selection (taking this further by using thousands of markers across an animal's entire
genome to produce more and more accurate calculations of the animal's
"estimated breeding value"); transgenics or genetic modification (transfer
of genetic material into an animal); and cloning (transfer of nuclear DNA

from a donor cell into an egg cell that has had its nuclear DNA removed) (see Twine 14–16). This developing field of eugenics experimentation also relies on methods that, only a few decades ago, were referred to as "new reproductive technologies," including freezing of sperm and eggs, artificial insemination, multiple ovulation, embryo transfer, and ultrasonography (95–96).[7]

With the use of these and other technologies (such as CT scanning and mid-infrared spectrometry), animal bodies are reduced to "factories for the production of protein for human consumption" (94), and the Cartesian *bête machine*, it seems, is fully realized. Individually and as a group, the technologies Twine discusses are designed to reduce the uncertainty that plagued traditional selective breeding practices, and to improve the profitability of the commercial product, whether eggs, milk, or meat: hence, for example, the use of CT scanning to predict carcass yield, transgenics to improve production traits, and genomic selection to identify points of genetic variation that might be used to select for different types of animal (15). Twine points out that the chicken genome was sequenced in 2004, the bovine in 2009, and that of the pig is imminent—developments contributing to apprehension of the animal body as a factory and/or as pieces of genetic information (15; 90–91). Indeed, he notes, the research of today's animal scientists involves "less and less lab-based work and more time in front of a computer screen doing work on database molecular information representation of animal bodies" (91). And where a body can be decontextualized as genetic information, or described as biotechnology, not only the practices, but also the discourses of, and ways of thinking about, animal breeding are undergoing radical transformation.

This is one point made by British geographer Lewis Holloway and his colleagues in their application to livestock breeding of Michel Foucault's concept of "biopower": the kind of power that emerged during and after the seventeenth century, to a large extent supplanting power that was localized in the figures of prince, pope, or police, with a more pervasive, and thus insidious, kind of power that was increasingly directed to the "administration of bodies" and "calculated management of life" (Foucault *History*, 140). Biopower permeates the individual and social body, proliferating new discursive formations, systems of knowledge, and (pedagogical, penal, medical and military) institutions, all bent on penetrating and disciplining bodies—so that, for Foucault, as Holloway notes, modernity's knowledge, and its power, do not belong to separate spheres.

Like Holloway, Twine also draws on Foucault's analysis of biopower as inclusive of practices (applied genetic and molecular technologies) and ideas, ways of thinking, speaking, and writing about these methods, as about animals, animal bodies, and human–nonhuman animal relations. This approach does not exempt animal ethics from the realm of bio-power—and it certainly encourages a critical assessment of the theoretical and ethical frameworks informing animal studies. Particularly now, Twine contends, when we know that factory farming is playing a major role in the climate change scenario that "threatens to radically change the material conditions for all life on the planet," it would be "irresponsible to refrain from casting a critical perspective over the tacit 'ethics' that support such an institutionalized and naturalized part of our global human-animal relations" (19). More than overt arguments in favor of factory farming, he has in mind here "the partial fiction of an 'animal ethics' which continues to speak to an obscuring human–animal dualism" (19). His argument is that the biotechnological vision of human-animal relations is being secured, rather than contested, by the reproduction of a "shallow" consideration of animal ethics (21).

A shallow animal ethics, for Twine, is one that does not engage critically with its own informing concepts, particularly if these perpetuate anthropocentric values and a dualist ontology. A shallow animal ethics presents itself as an autonomous subfield of knowledge, rather than opening to a diversity of approaches, philosophical and otherwise, and to the pluralistic interdisciplinarity that increasingly characterizes animal studies. Animal ethics, he contends, "cannot be fenced into a narrow philosophical discourse" (19). Reviewing the key ethical arguments that have been constitutive both of animal ethics and initially of animal studies itself, Twine points to two criticisms of Singer and Regan: that they have "put too much onus on similarity defined in human terms," and that "they remain uncritically tied to a liberal humanist notion of the subject defined as an independent autonomous individual" (28). The key word here is *uncritically*: Regan and Singer do not call their informing concepts into question. For Twine, theirs, then, is a shallow animal ethics, theorized "via an overly cognitive and rather disembodied ontology" (29). For all of their opposition to Descartes, they nonetheless reconstitute a version of the essentially rational Cartesian subject (29).

5

THE DEATH OF "THE ANIMAL"

What's in a Word?

The English word *animal* comes from the Latin *animalis*: having the breath of life, animate, from *anima*, life, breath. Citing this Latin origin, the *Oxford English Dictionary* defines "animal" as a living creature, anything living or having the breath of life—a broadly inclusive definition, certainly of humans. Yet, the OED goes on to list several, more narrow, definitions of the word, among them: one that distinguishes humans from animals ("one of the lower animals; a brute, or beast, as distinguished from man"); one that distinguishes animals from plants ("a member of the higher of the two series of organized beings, of which the typical forms are endowed with life, sensation, and voluntary motion, but of which the lowest forms are hardly distinguishable from the lowest vegetable forms by any more certain marks than their evident relationship to other animal forms, and thus to the animal series as a whole rather than to the vegetable series"); and still another that defines "animal" as a term used "contemptuously or humorously for a human being who is no better than a brute, or whose animal nature has the ascendancy over his reason; a mere animal" (OED Vol. 1, 333). Not surprisingly, perhaps, dictionaries betray the binary oppositions on which a culture relies. *Webster's Dictionary*, another case in point, provides its own several definitions of "animal": the first applying the word to all multicellular organisms, including even protozoa and other single-celled eukaryotes that have motility and the capacity to acquire and digest food; the second defining "animal" as a living thing other than a human being; a third stating that the word is applicable to a "mammal as opposed to a fish, bird, etc."; a fourth defining "animal" as "the physical, sensual, or carnal nature of human beings; animality: *the animal in every person*"; and a fifth entry applying the word to "an inhuman person, a brutish or beastlike person: *She married an animal.*" To make what are two obvious points: first, such dictionary definitions are not so much "scientific" or "factual" as reflections of historical and contemporary prejudices that concern us in this book; and second, the word *animal*, as illustrated by the above definitions, gathers a multiplicity of different species and members of species—whether "living creatures," "multicellular organisms," "mammals," or "beastlike persons"—into the sameness of a single word, in this way blurring differences in favor of one unifying term, something that happens, by the way, every time a binary opposition is set up.

This chapter considers the work of two philosophers who put "the animal" abstraction into question, Italian analytic philosopher Paola Cavalieri, editor of the Milan-based quarterly journal *Ethics and Animali*, and Algerian born French thinker, Jacques Derrida, one of the twentieth-century's major continental philosophers. Both locate "the animal" within the tradition of Western metaphysics, but out of two different understandings of what this tradition entails. By way of focusing on Cavalieri's *The Death of the Animal*, and Derrida's *The Animal That Therefore I Am*, the chapter aims to open questions about "the animal" at the center of animal studies—questions directed to making animal studies a more *critical* enterprise.

The Death of the Animal

Among those who are concerned today with animal welfare issues, ecology, biodiversity, and survival of the planet, most, if not all, would support Paola Cavalieri's contentions in her 2001 book, *The Animal Question: Why Nonhuman Animals Deserve Human Rights* that moral consideration is owed to nonhuman animals as well as to humans, and that the principle of equality need not presume that no differences exist between or within sexes, or races—or animals—considered as groups; in other words, that moral equality does not depend on empirical equality (5). We would not find the same consensus on Cavalieri's argument that the doctrine of human rights is particularly well suited both to overcoming the historical human/animal, mind/body, hierarchy—opening a door to the possibility that equality does not belong to *Homo sapiens* alone—and, in the process, bringing about "the death of the animal," an expression with which she alludes to ridding philosophy of "the animal" abstraction.

In *The Animal Question*, Cavalieri calls for an "expanded" theory of human rights, one that includes within the sphere of those having full moral status all "intentional beings," by which she means all beings that have goals and want to achieve them (137). She insists that this "intentionality" criterion is free of reference to capacities such as rationality and self-consciousness that are invariably at issue when hierarchies are established. Indeed, in an expanded theory of human rights, she glimpses the promise of eliminating historical hierarchies altogether, therefore ridding us of the reductionist concept of "the animal" that grounds the elevation of humans over animals, the human/animal binary. She suggests in this book that, while the question of which nonhuman animals "meet the requisites for inclusion in the privileged area of full moral status" cannot yet be determined in detail, "among the beings that an *expanded theory of human rights* should cover there undoubtedly are mammals and birds, and probably vertebrates in general" (140). Presumably, if these species were included in the "privileged" moral sphere, the human/animal dichotomy would disappear: hence, the title of her 2009 book, *The Death of the Animal*, the study that concerns us here.

In a format that Cavalieri has adopted before, *The Death of the Animal* proceeds as a "Socratic" dialogue. The form is a creation not of Socrates, but of his student Plato. Tradition has it that Socrates wrote nothing at all, leaving Plato, for all of his disparagement of writing, to become the "writer," the one who transcribed his teacher's speech. Plato did this by creating dialogues in which Socrates appears as a character, a wise teacher, who instructs his interlocutor in the ways of truth. It is a curious form for Cavalieri to adopt in that, for Plato, the founder of Western metaphysics, Socrates' instructions invariably put in place the either/or binary structure out of which metaphysics emerged, the fundamental intelligible/sensible binary in its various forms, for example, in the hierarchies *logos/mythos*, philosophy/literature, male/female, and human/animal.[1] In keeping with Plato's dialogic form, Cavalieri has Alexandra Warnock, one of the participants in *The Death of the Animal*, assume a role much like that of Socrates. Not incidentally, Alexandra is an analytic philosopher, as is Cavalieri herself. Alexandra's interlocutor, Theo Glucksman, is a continental philosopher, much less wise, and less critically and ethically astute, than his analytic teacher. Poor Theo, we learn as the dialogue opens, does not even know the meaning of "moral status" (3)! Appropriately, the dialogue takes place on an island in Greece.

Cavalieri's dialogue centers on a critique of the view she has Alexandra refer to as "perfectionism," the view, present in many Western approaches to ethics, that one's moral status, how much one counts, depends on the degree to which one possesses certain favored characteristics. Extending back to Aristotle, perfectionism has historically made use of "a specific philosophical abstraction," Alexandra explains to Theo: "'The animal' is what lies at the bottom of the perfectionist's hierarchy" (3). "The animal," as "par excellence, the negative term of comparison," has facilitated the overall derogation of others: "Historically, the subjugation of human beings has been usually coupled with their 'animalization'—think of slaves, women, the disabled, native peoples . . ." (3–4). The time has come, says Alexandra, to erase the philosophical abstraction, "the animal," this negative term, from our mental landscape (4). She is careful to point out that "the animal" belongs to metaphysics; it is a metaphysical concept (4).

In arguing against metaphysics from within its inaugural, dialogical, form, Cavalieri might seem to be using the tradition against itself. At the same time, however, even before the dialogue begins, she has Alexandra espousing a version of the *logos/mythos* binary, in this case, through the claim that the "normative" or "prescriptive" sphere of ethics must hold itself apart from all things "descriptive."[2] As Alexandra clarifies on the First Day, descriptive discourses (including myths, literary narratives, even philosophical master–narratives) should not to be translated into the domain of normative ethics, for they cannot "aspire to receive a general and uncontroversial rational assent" (7). They are not "verifiable," not subject to proof, and thus, according to Alexandra, they are "metaphysical" (7). For example, she cites Aldo Leopold's view of the integrity and beauty of the biotic whole, which she says cannot be used to guide the conclusions of normative philosophy (8). For Cavalieri, as she has her two interlocutors make clear, all descriptive systems, Martin Heidegger's philosophy as another example she gives, are both *archaic*, an "unquestioned legacy of the past," mostly of earlier religious systems, and *dogmatic*, in the sense of resting on "unproved principles or facts" (9). As well, they tend to translate into forms of perfectionism, that is, into hierarchies of moral status: for example, Heidegger's discourse on the "world-poverty" of nonhuman animals (9; 11).

Offering additional clarification of her division of prescription from description, Cavalieri has Alexandra explain to Theo that philosophical ethics itself divides in two. In the broadest sense, it is an all-inclusive endeavor that includes precepts about the good life, character traits to be fostered and values to be pursued. Descriptive systems may find some room within this broad-based sort of ethics. But they are not acceptable within what Alexandra calls "narrow morality," that is, normative or prescriptive morality, the kind of morality that interests her, and that has to do with negative constraints on conduct, such as do not harm or do not kill (9). Narrow morality concerns itself not only with moral agents— where a moral agent is defined as "a being who can reflect morally on how to act, and who can be held accountable for her actions"—but also with moral patients, that is, with "beings whose interests deserve respect" (13). Adopting here a version of Tom Regan's distinction between moral agents and moral patients, Cavalieri has Alexandra explain that moral agents may become moral patients when they are on the recipient end of an action, but generally, moral patients are those cast on the lower end of perfectionist hierarchies—such as disabled humans, handicapped infants, and animals—beings who are vulnerable because they lack the usual criteria required for full moral status: rationality, self-consciousness, and conceptual-linguistic capacities (16–22). None of the ethical approaches she reviews for Theo (virtue ethics, theories of intrinsic value, contract theory, bioethics) is free of the perfectionist tendency to overlook moral patients. Only normative ethics of the analytic method, "marked by clarity and explicit argumentation" (38), proves sufficient, Alexandra explains to Theo, who, for Cavalieri, plays the role of a docile student, his continental philosophy dismissed as "metaphysical" throughout the dialogue.

As in her earlier book, *The Animal Question*, so in *The Death of the Animal*, Cavalieri appeals, within the analytic tradition, to human rights theory as promising the end of perfectionist hierarchies, and so "the death of the animal" (40). As the dialogue concludes, she has Alexandra summarize the chief merits of human rights theory: (1) it allows for the autonomy of normative ethics, the kind of ethics that bases itself on "uncontroversial rational assent," that is marked by clarity, and by prescriptions based on fact and subject to proof; (2) it protects (especially vulnerable) individuals from interference by stressing the fundamental negative rights to life, freedom, and welfare; (3) it requires, for access to the sphere of rights holders, only intentionality—the intentionality of a being that has goals and wants to achieve them (39). For those who qualify as rights holders, it is a radically egalitarian theory, Alexandra insists, and one that moves equality from the descriptive to the normative sphere. Which nonhuman animals can be considered intentional and thus holders of human rights is not taken up in this particular book.

At least two terms are centrally at issue in Cavalieri's *The Death of the Animal*: one is "the animal" and the other is "ethics." The importance of the latter is clear to Peter Singer in his Foreword to the book, where he suggests that the dialogue concerning "the death of the animal" is itself a "launching pad" for the book's real debate: "a distinct and very lively debate about the nature of philosophy and the role that reason can play in ethics" (x). That debate comes down to analytic versus continental philosophy, as is suggested by the two traditions from which the dialogue's protagonists come (x). As Singer puts it, the dialogue illustrates "the clarity of analytic philosophy" (x) contra such continental thinkers as Heidegger and Derrida, and, he says, it demonstrates that the same analytic philosophy that ushered in the current animal movement alone offers the kind of ethical reasoning that can take the movement forward from here. In other words, Singer supports Cavalieri's dismissal, through Alexandra, of "continental philosophy," apparently in its entirety, as if collapsing the multiple variants of that tradition under a single label were not an instance of the kind of abstraction that the "the animal" represents. As Matthew Calarco puts this point in the "Roundtable" that is published in *The Death of the Animal* following the dialogue, the term "continental philosophy" is "little more than a family resemblance concept that gathers together a broad range of thinkers and texts" (74).[3] The attempt to exclude all of these thinkers and texts from animal ethics is problematic, particularly since, in Calarco's view, the criticisms that Cavalieri makes against the continental tradition "seriously miss the mark" (75).

For example, in arguing for the autonomy of ethics, "the view that ethics is a theoretical inquiry endowed with its own standards of justification, within which criteria coming from other domains—be they religious, metaphysical, or scientific—have no direct relevance," Cavalieri (*Death*, 38) charges continental philosophers with having subordinated ethics to metaphysics and to its anthropocentric hierarchies. Or, as Singer summarizes this point in his Introduction to *In Defense of Animals: The Second Wave*, referring to Cavalieri's essay in that volume, her "contrast between the human-centered approaches taken by Heidegger and Derrida and the more egalitarian approach taken by many English-language [*sic!*] philosophers reveals the conventional self-interest that often lurks behind what appears to be deep metaphysics" (7). Matthew Calarco finds this criticism to be "deeply uncharitable and inaccurate" in that nearly all of the continental philosophers Cavalieri discusses in her work (including Nietzsche, Heidegger, and Derrida) "do *not* subordinate ethics to metaphysics," or subscribe to "a hierarchical and anthropocentric ethics (at least not in any straightforward manner)" ("Toward," 75). Quite the contrary, what motivates the work of the continental thinkers mentioned above "is trying to contest Western metaphysics and create the possibility of developing a mode of thought and practice that does not fall back squarely within the conceptual and practical constraints of the metaphysical tradition" (76). Calarco explains that these thinkers all approach ethics and the question of the animal outside the framework of rights—precisely because "they believe (and rightly so, in my opinion) that *rights discourse and its axioms are themselves metaphysical*" (76).

Another question Calarco raises, one I have mentioned in earlier pages, is the question of moral status: like nearly every philosopher working in the analytic tradition of animal ethics, he notes, Cavalieri seeks to determine criteria that determine who counts or who belongs to the moral community, an idea that strikes him as "deeply metaphysical and subordinated to traditional metaphysical ends" (77). Indeed, for Calarco, "the effort to determine who does and does not belong to the moral community is one of the most problematic foundational gestures in the Western metaphysical tradition and is indicative of its imperialistic tendencies" (77). Rather than the question of who belongs, Calarco asks this: "What if the logic of deciding inclusion and exclusion within the moral community were itself the problem? And what would be the implications for ethics if we were to abandon the aim of determining the proper limits of moral status altogether?" (77). These are very different questions from the "who counts?" debate that now preoccupies analytic animal ethics. They are crucial questions, and ones to which we will return.

What's in a Word?

In 2008, four years after his death, Fordham University Press published Jacques Derrida's *The Animal That Therefore I Am*. The book, edited by Marie-Louise Mallet and translated by David Wills, collects the papers Derrida presented at a ten-day conference on his work held in 1997 in France, and titled "The Autobiographical Animal." Born in 1930 in Algeria, where he lived until moving to France in 1949, Derrida was one of the first twentieth-century philosophers to call concerted attention to "the animal" question—which for him is "not just one question among others," as he remarks in "Violence Against Animals," an interview with Elisabeth Roudinesco, but "the limit upon which all the great questions are formed and determined" (62–63). From his first publications, Derrida took the question of animality—the thinking of human and non-human animal life and of human and nonhuman animal relations—to be "decisive" in this sense, and for this reason, he goes on to say in the interview with Roudinesco, he attempted to extend his thinking of the *trace*, elaborated in such early texts as *Of Grammatology* "to the entire field of the living" (63). All of Derrida's key concepts represent his attempts to exceed anthropological limits—*différance* for another example, a term he discusses in a 1968 essay under that title—and animality is an insistent theme across the entire spectrum of his work: from "Plato's Pharmacy," through "White Mythology," the *Geschlecht* papers, *Of Spirit*, "A Silk-worm of One's Own," *Rogues*, and two volumes of *The Beast and the Sovereign*.[4]

To some extent, the concerns that Derrida raises in *The Animal That Therefore I Am* are quite familiar within the discourse of animal studies, and they are concerns that have been brought forward in earlier pages of this book. For example, in line with what we have referred to as the *disappearance* of animals, Derrida writes that over the course of the last two centuries in particular, traditional forms of the treatment of animals have been turned upside down by developments in zoological, etho-logical, biological, and genetic forms of knowledge and, as inseparable from these, by "*techniques* of intervention" that have actually transformed the animals that are their targets (25). This transformation has occurred through genetic experimentation, massive scale artificial insemination, industrialized farming, and in general, through the reduction of ani-mals "not only to production and overactive reproduction (hormones, genetic crossbreeding, cloning, etc.) of meat for consumption, but also of all sorts of other end products" (25). This is all well known, Derrida adds; no one can today deny "the *unprecedented* proportions of this sub-jection of the animal" (25). He refers to "animal genocides: the number of species endangered because of man takes one's breath away" (26). And again, he suggests that all of this is "well known," that everybody knows what "terrifying and intolerable pictures a realist painting could give to the industrial, mechanical, chemical, hormonal, and genetic violence to which man has been submitting animal life for the past two centuries. Everybody knows what the production, breeding, transport, and slaughter of these animals has become" (26).

However, Derrida's modes of contending with this unprecedented subjection of animal life are not so familiar within animal studies, in part for reason of attempts to exclude his work from the field—attempts such as those made by Singer and Cavalieri—because it does not constitute either a utilitarian or rights-based Anglo-American analytic philosophy. Moreover, to elaborate on a point Matthew Calarco makes in *The Death of the Animal* Roundtable, Derrida calls for a rethinking of the history and concept of "rights," which belongs to metaphysics and which "until now, in its very constitution, has presumed the subjection, without respect, of the animal" (*Animal*, 87). Thus, while indicating his "strong sympathy" for the *Universal Declaration of Animal Rights*, Derrida suggests that the modern discourse of rights depends on a concept of the subject that comes down to us from Descartes' "I think" or "I am"—a concept that not only dominates post-Cartesian philosophy but that, in our world, in "modern times," is "the discourse of domination itself" (88–89). Importantly, Derrida suggests that human domination over animals is exercised not only "through the boundless wrong we inflict on animals," but also through "forms of protest that at bottom share the axioms and founding concepts in whose name the violence is exercised" (89), in other words, through declarations of, or ethical arguments for, "animal rights," both of which can be forms of what, in the previous chapter, Richard Twine, following Michel Foucault, calls "biopower."[5]

Rather than adopting, without question, the founding concepts that have for centuries facilitated human domination over animals, Derrida engages in critical questioning of them, which is why in *The Animal That Therefore I Am*, for example, he indicates that today's subjection of animals is a *history* he is "attempting to interpret" (25). In this book alone, this work of inheriting tradition includes readings of Genesis, Plato, Aristotle, Descartes, Kant, Hegel, Nietzsche, Heidegger, Freud, Levinas, and Lacan, among others: meticulous analyses of the conceptual frameworks that are bequeathed to us by the tradition of metaphysics, *critical* questioning that is necessary to an understanding of the many ways in which, in the Western tradition, which is never homogeneous, the border between man and animal has been determined.[6] Throughout the book, Derrida turns several times to "the animal," the abstraction that also concerns

Cavalieri—although in important and instructive ways, his analysis differs markedly from hers.

What's in a word? "The animal is a word, it is an appellation that men have instituted, a name they have given themselves the right and authority to give to the living other" (Derrida *Animal*, 23). In the first place, the word belongs to the West's logic of inclusion/exclusion: it is, and has been, used to separate *Homo sapiens* hierarchically from "the living other." Traditionally, as Derrida demonstrates in *The Animal That Therefore I Am* and in a number of other texts, the delineation of this hierarchical logic of inclusion/exclusion has been based on various powers or capacities that man accords to himself but denies to "the animal," characterizing the latter by an incapacity for the *logos*, for speech, hearing, and response (rather than the reaction of an automaton); an incapacity for friendship, family, death mourning, and even, as Jacques Lacan has it, for pretending to pretend. Insofar as contemporary analytic philosophers continue to ground ethics on this same capacities argument, they can only reinstate the inclusion/exclusion, same/different, man/animal hierarchy.

"*Animal* is a word that men have given themselves the right to give" (32). Although, as Derrida notes, "[t]here is no Animal in the general singular, separated from man by a single, indivisible limit" (47), men have given themselves this word precisely in order to assimilate differences into sameness, or as he puts it, "in order to corral a large number of living beings within a single concept" (32). This is a second reason why, for him, "a certain wrong or evil" (32) derives from the word, which, supposedly representing a kind of unity, serves as a cage, a point we will return to in the next chapter. "Confined within this catch-all concept, within this vast encampment of the animal, in this general singular, within the strict enclosure of this definite article ('the Animal' and not 'animals')," Derrida writes, "as in a virgin forest, a zoo, a hunting or fishing ground, a paddock or an abattoir, a space of domestication, are *all the living things* that man does not recognize as his fellows, his neighbors, or his brothers" (34). Calling attention to this problem, Derrida suggests a *serious* "word play": as an alternative to "the animal," he coins the term *l'animot*. In the French language in which he wrote, *l'animot* is homonymous with *l'animaux*, animals in the plural: the singular word inscribes not one "animal" but a multiplicity of "animals." *L'animot* also has the word *word* (*mot*), embedded in it, which reminds us not only of what is at stake in a word, and in the historical exclusion of animals because they lack words, it also reminds us that ethics itself is caught up in language, and therefore cannot lay claim to the kind of self-presence or mastery that it often denies to living others. Finally, *l'animot* interests Derrida as "a chimerical word," a "sort of monstrous hybrid" (41) that contravenes the laws of the French language and that recalls for him a number of mythological animal-animal and animal-human hybrids, special creatures that are anything but mono-types.

Ethics

As we have noted, the dialogue Paola Cavalieri sets up in *The Death of the Animal* uses "the animal" abstraction as but a "launching pad" (Singer) for her argument concerning the autonomy and superiority of a rights-based prescriptive analytic ethics over all other approaches to ethics, particularly those of continental philosophy. In line with Peter Singer and Tom Regan, Cavalieri foregrounds the "who counts?" question as central to ethics, advocating "intentionality" as the requirement for inclusion in the sphere of moral rights. For Derrida, too, the meaning of "ethics" is much at stake in the word "animal"— although unlike Singer, Regan, and Cavalieri, he does not conceive of ethics as a logic of inclusion/exclusion, a discourse that, by way of a "who counts?" calculus, bases human/animal difference on "a power that consists in having such and such a faculty, thus such and such a capability, as an essential attribute" (Derrida *Animal*, 27). Rather, as Calarco suggests, Derrida takes this logic itself to be the ethical problem. Along with "the animal," we might say, "ethics" for Derrida is a word that men have given themselves the right to give to the discourse through which they declare who does and who does not possess the "*can-have* [pouvoir-avoir] of the *logos*" (27), the power or capacity that is taken to be prerequisite for moral worth. For Derrida, such an ethics falls squarely within the tradition of metaphysics, in that it determines human/animal difference on the basis of some animal deprivation, "the animal deprived of the *logos*, deprived of the *can-have-the-logos*: this is the thesis, position, or presupposition from Aristotle to Heidegger, from Descartes to Kant, Levinas, and Lacan" (27). And, despite its nonspeciesist claims, such an ethics is human-centered and "self-interested" in that its "*can-have*" is based on powers or capacities that man sees (misrecognizes) in himself.

According to Derrida, Kelly Oliver points out, "[d]elineating rights, calculating interests, and weighing the value of one life against another may be juridical necessities in civil society, but they are the antithesis of ethics" (36). Moreover for Derrida, Oliver notes, conceiving of ethics as normative or prescriptive "preprograms" moral decision-making, reducing it to the automatic following of moral principles or rules: not a response at all, but a reaction that in the end "belies the Cartesian distinction between reacting and responding supposedly definitive of the man/ animal distinction" (37). As Cary Wolfe puts it in the Roundtable discussion included in Cavalieri's *The Death of the Animal*, "by relying upon a one-size-fits-all formula for conduct," an ethics that concerns itself with calculating interests or rights "actually *relieves* us of ethical responsibility" in favor of "an application that, in principle, could be carried out by a machine" ("Humanist," 53). And to return to a point made in the above paragraph, the "most like us" standard on which these calculations are based attributes illusory power to the human subject – a subject that Wolfe refers to as a "fantasy figure of *the human*," a subject "in denial of its own vulnerability" (55). To invest in this subject is to imagine "that we *can* calculate the incalculable, that we *can* decide the undecidable, that we can be *certain* about what is just, fair, equal or right" (Oliver *Lessons*, 36).

Even suffering is considered a power or capacity within the tradition of analytic ethics, a *can-have* attribute of the *logos*. Thus Peter Singer writes in his Introduction to *In Defense of Animals* that: "Although nonhuman animals certainly can grieve for the loss of those to whom they are close, the nature of the grief must differ in accordance with the differing mental capacities of the beings" (6). Derrida receives Jeremy Bentham's dictum—that the question is not to know whether animals think, reason, or speak, but whether they suffer—very differently, suggesting that "the form of this question changes everything," in that after this question, the philosophical problematic of "the animal" can no longer center on the *logos*, mental capacity as in the power to reason, to speak, "and everything that that implies" (27). For Bentham's question, "is disturbed by a certain *passivity*," Derrida writes; it already manifests, as a question, "the response that testifies to a sufferance, a passion, a not-being-able" (27). At the heart of this passage is Derrida's thinking of the *trace* that marks the finitude humans share with animals: the trace that, prior to any opposing of activity to passivity or of sameness to difference, is constitutive of the self as other.

Passivity suggests an instance of exposure to (the suffering of) the other in which the self, in a radical suspension of self-certainty and self-presence, suffers itself as other.[7] Passivity is not a power or capacity, for

Derrida, but an *impouvoir*, a "nonpower at the heart of power" (28), a heightened sense of one's vulnerability before—and incapacity, finally, *to know*—another animal. Passivity is not the frontal posture of a subject who deals head-on with the problem of who is qualified to count as his or her fellow (see Derrida's "Passions," 10–11). Rather, it is destabilizing brought by seeing oneself being seen through the eyes of the other, through the eyes of an animal, an experience through which the self, stripped of the mantle of autonomous first-person author, is called on as an addressee, as one who comes after: "*To follow* and *to be after* will not only be the question, and the question of what we call the animal," Derrida writes in *The Animal That Therefore I Am*. The French title of the book (*L'animal que donc je suis*), while playing on the Cartesian "I am," conjugates more than one verb: *être*, to be, and *suivre*, to follow, so that *je suis* means both "I am" and "I follow," one as inseparable from the other: ethically, to be human is to follow, to be positioned not as a first-person speaking subject, but as an addressee. And in the title, as in the book, everything links being-in-the-world-as-following to the question of "what *to respond* means" (10). For Derrida, "I am" insofar as "I follow": both in the sense that *I come after* and am called on to respond, and in the sense that the trace, or what he calls *différance*, precedes the authorial "I"—the I who constructs the same/different, human/animal opposition. It is the trace that, in Derrida's thinking, undermines modernity's life/death, animate/inanimate bifurcation.

In brief, and to close this chapter, Derrida refers to the passivity that disturbs Bentham's question—"Can they suffer?"—as "a possibility without power, a possibility of the impossible" (28), and for him, morality, in each and every case, "resides there, as the most radical means of thinking the finitude that we share with animals, the mortality that belongs to the very finitude of life, to the experience of compassion, to the possibility of sharing the possibility of this nonpower, the possibility of this impossibility, the anguish of this vulnerability, and the vulnerability of this anguish" (28). He holds out passivity as an experience of the impossible, and at the same time, as the possibility of an ethics that exceeds, goes beyond, calculation and preprogrammed rules; as an ordeal that an ethical ethics must suffer in every response to the "wholly other they call 'animal'" (13).

6

OPEN THE CAGES

Animots and Zoos

In undergraduate university classrooms, "zoo" is popular shorthand for zoology studies and/or for various "zoo" sciences, such as zoophysiology (the physiology of animals, as distinct from humans), zoonosology (the study and classification of animal diseases), and zoochemistry (the biochemistry of solids and fluids in animal bodies), all of these "zoo" specializations premised on fundamental differences between human and animal life. The word "zoo" can also be used pejoratively to refer to a disorderly, haphazard place and/or to a bizarre group of people there assembled, as in "This pub is a zoo on Friday afternoons." Similarly, "zooey," in American slang, is an adjective signifying "wild" or "barbaric" (Green 1311). The word zoo, as commonly used to refer to zoological parks or gardens, such as the Zoological Gardens in Regent's Park, London, informs all of these popularizations, for historically, zoos have separated ("wild") animals from humans, often by using bars and cages, enclosures that, particularly in early zoos, led the trapped animals to exhibit neurotic ("bizarre") behavior, displaying, as it were, their profound difference from human spectators assembled to view the display.

The first modern zoos—founded in Vienna, Madrid, and Paris in the eighteenth century, and London in the nineteenth century—housed animals in cramped and unsanitary conditions, so that, as Lori Gruen remarks, "though visitors were amazed and amused by what they saw, they often complained about terrible smells, flies and filth in cages, and lethargic animals" (137). Into the nineteenth century, zoos came increasingly to typify what biologist David Ehrenfeld calls the "Age of Control," with its themes of "exploration, domination, machismo, exhibitionism, assertion of superiority, manipulation" (xvii). Changes (e.g., enlarged enclosures redesigned to replicate natural habitats) were made slowly during the twentieth century, as research and education gradually replaced entertainment as the central justification for zoos (Gruen 138). Conservation, in part through captive breeding programs, has also emerged as a *raison d'être* for twenty-first-century zoos. According to some, then, as Ehrenfeld points out, zoos are now playing a crucial role "in conserving the vanishing species of the world's ecosystems, and in educating children and adults about the value and plight of nature" (xviii). Still, according to others, zoos "are conserving little or nothing" and are only "teaching and reinforcing destructive ways of relating to nature," an argument according to which "society would be better off if zoos became extinct" (xviii).

Ehrenfeld's comments come in his Foreword to *Ethics on the Ark: Zoos, Animal Welfare, and Wildlife Conservation* (edited by Bryan Norton et al.), a collection of essays on zoos that represents both opinions, both sides of the debate that Ehrenfeld characterizes as either "pro-zoo" or "pro-animal rights" (xix; see also Malamud). His labeling is apt, in that most ethical arguments against zoos emerge from within the analytic tradition and its calculus of interests or rights. Thus, for example, Dale Jamieson maintains that zoos "teach us a false sense of our place in the natural order. The means of confinement mark a difference between humans and other animals" (142). In making this case, Jamieson adopts a version of Peter Singer's "like interests" approach, suggesting that, "if everything else is equal," we should respect animals' as well as humans' interest in liberty (133–134). The problem, as we have suggested, is that both the "like interests" and "rights" approaches involve an inclusion/exclusion calculation that reinstates the logic of human/animal difference Jamieson deplores, and that grants moral status only to animals "most like us." In *Animals Like Us*, Mark Rowlands, following Jamieson, argues that zoos are "morally illegitimate, and should be abolished" (159) because they cut deeply into an animal's autonomy, where autonomy (a human-based standard) is one of the most vital of all "interests" (153–154). Gruen, too, takes up an interests calculation: in order to determine whether it is wrong to keep animals well-provided for but in captivity, we need to know whether animals other than humans have an interest in liberty as such, she says, an interest that zoos violate; we need to know whether "any other animals [are] the sorts of beings who can be said to value liberty intrinsically" (144). For Robert Garner, zoos constitute an illegitimate infringement of animal rights (142). This aligns with Tom Regan's argument that, given the rights view, keeping animals in zoos is wrong (see his "Are Zoos Morally Defensible?" 44–45), for as he puts it in *Empty Cages*: animals (mentally normal mammals of a year or more), who are just like us in being subjects-of-a-life, possess the right not to have their liberty infringed upon (58–61). Hence, "we must empty the cages, not make them larger" (61).

This chapter surveys some other sorts of "zoo" studies, all of which have emerged out of engagements with continental philosophy or out of work broadly labeled *posthumanist* (a term we define below), studies that question the adequacy of a "rights" or "interests" framework for animal ethics and for contending with the legacy of the West's human/ animal hierarchy. Characterizing these approaches overall is a comprehensive critique of what we have referred to as the *fantasy*, what Derrida calls the *phantasm*, of the autonomous self: the idealist account of the self or "subject" that prevails in metaphysics, that casts the subject as somehow transcendent to language and as author of ethics as a prescriptive "science" marked by clarity and rooted in fact. The work discussed below asks "what it would mean in both intellectual and ethical terms to take seriously the question of the animal—or the *animals*, plural" (Wolfe *Animal*, 190), *l'animots*. It attempts to open what Derrida calls the cage or "encampment" of the word "animal," and along with that, to move beyond the enclosing concepts of "the human" (the "just like us humans" standard), "the subject" (or "subjects-of-a-life"), "rights," and "ethics" as a preprogrammed calculus.

Zoontologies

Cary Wolfe opens the Introduction to his edited volume, *Zoontologies: The Question of the Animal*, with an extensive list of leading writers of the past three decades who have devoted considerable attention to the question of "the animal" under a variety of figures and themes. The list, which he says could easily be extended, includes the names of Julia Kristeva, Jacques Derrida, Gilles Deleuze and Félix Guattari, Jacques Lacan, Slavoj Žižek, Stanley Cavell, Georges Bataille, René Girard, bell hooks, Michael

Taussig, Étienne Balibar, Donna Haraway, Katherine Hayles, and Evelyn Fox Keller. Distinguishing all of this work, Wolfe suggests, is its critique of, and through this critique, its attempted movement beyond, traditional humanism, particularly humanism's foundational "subject" or "self." Herein lies a major difference of this "posthumanist" work from animal rights philosophy, one of the central ironies of the latter being "that its philosophical frame remains an essentially humanist one in its most important philosophers (Peter Singer and Tom Regan), thus effacing the very difference of the animal other that animal rights sought to respect in the first place" (xii).[1] According to Wolfe, posthumanist studies are nonidealist approaches that, rather than investing in some version of Cartesian rationality or mental capacity, take for granted the embodiment and "specific materiality and multiplicity of the subject" (xiii). In other words, such studies take seriously the human-as-animal, which is implicit in the title of Wolfe's book, combining the prefix "zoo" with "ontology" (thinking or philosophy concerned with the nature of being). Wolfe's collection of essays evidences the wide range of zoontological work currently underway, from scholarly theories to popular culture contributions. Such plurality characterizes posthumanist animal studies: there is no interest here in establishing the hegemony of philosophy, or of one branch of philosophy, over feminism, biology, psychoanalysis, anthropology, fiction, poetry, religious studies, cultural studies, and so on.[2] And although not all of the theorists Wolfe lists in his Introduction consider the "post" (whether posthumanism, postmodernism, or poststructuralism) an unproblematic term—insofar as, for one thing, in implying what comes "after," it is bound up with a linear model of history and time—Wolfe does not use the term to suggest linearity, historical ruptures, or historical-theoretical unities: his attempt is to explore the thinking of multiplicities.[3]

One point Wolfe makes in *Zoontologies* concerns the importance of Sigmund Freud's work for contemporary studies of animality and of the legacy of human/animal difference. Indeed, he writes, "it is scarcely possible to think about what the animal means to us in the modern and postmodern period without working through Freud's theories of drive and desire and the anthropological work on sacrifice and sexuality of *Totem and Taboo* and *Civilization and Its Discontents*" (xiii–xiv). His comment suggests another way in which the approaches he considers differ from those of analytic philosophers, within whose work Freud is noticeably absent. As but one example, taken from Wolfe's collection, a collection that is impossible to summarize here, in "From Protista to DNA (and Back Again): Freud's Psychoanalysis of the Single-Celled Organism," professor of English Judith Roof considers Freud's interest in "the protist and its twin the 'germ plasm' [as] primal, deathless reference points" for his thinking about life processes in general, and since the single-celled organism both is and is not "human," for his thinking about life in a nonspeciesist way (101). In his contribution to the collection, "Animal Body, Inhuman Face," philosopher Alphonso Lingis turns also, through the "face," to a thinking of life and of species in a symbiotic way, non-speciesist way. As Cary Wolfe writes of the Lingis essay, "the thrust is toward opening the human to the heterogeneity and multiplicity within which it has always been embedded" (xix).

Again, considering "posthumanism" in his Introduction to *Animal Rites: American Culture, the Discourse of Species, and Posthumanist Theory,* Wolfe links the term to a recognition that "the 'human' is inextricably entwined as never before in material, technological, and informational networks of which it is not the master, and of which it is indeed in some radical sense 'merely' the product" (6). Broaching speciesism as integral to the formation of Western subjectivity and sociality, posthumanism acknowledges "that the full transcendence of the 'human' requires the sacrifice of the 'animal' and animalistic, which in turn makes possible a symbolic economy in which we can engage in what Derrida will call a 'noncriminal putting to death'" of the other, not only animals, but also humans marked as animals (6; see also chapter 10 in this volume). Wolfe titles his study *Animal Rites* (not *Rights*), then, because, as W. J. T. Mitchell, critical theorist and long-time editor of the journal *Critical Inquiry*, points out in his Foreword to the book, the humanist discourse of "rights" is itself predicated on the difference between humans and animals, and thus requires the drawing of included/excluded lines (ix). In the book, in Mitchell's words, "Wolfe works through a series of philosophers—Wittgenstein, Cavell, Lyotard, Deleuze and Guattari, Lévinas, and Derrida—who have radically reshaped the traditional view of 'the' animal as a straightforward antithesis and counterpart to 'the' human," suggesting along the way, "the possible emergence of a new, postmodern bestiary" (xii). Mitchell suggests further that *Animal Rites* must be read in the context of the current extensive outpouring of work on questions of culture and nature, the human sciences and biology: "The question of the animal is just one component in a rethinking of a whole set of nonhuman entities that seem to take on organic, lifelike, or 'auto-poietic' characteristics—intelligent machines, of course, but also systems and swarms, viruses and coevolutionary organisms, corpses, corpora, and corporations, images and works of art" (xiii). This is to say, in part, that when Wolfe talks about animals, "he is not thinking only of chimps and whether they speak to us. It is the nonlinguistic that matters just as much as the capabilities of the higher animals" (xiii).

Although Wolfe's collected essays, along with his own studies of animality, are "posthumanist," this does not mean that he or they discount the humanist tradition, but more along the lines of Derrida's under-standing, that "opening the cages" is first of all a work of inheritance, a task that requires critical engagement with the liberal humanist tradition, including contemporary utilitarian and animal rights philosophy. As he puts it in *Animal Rites*, the point is not to reinstate the "self-serving abstraction of the subject of freedom" (9). At stake in these studies is "a major reassessment of how distinctly different figures in contemporary philosophy and theory have thought about the question of language in relation to the difference between human and animal," this as involving a rethinking of "the relation between language, ethics, and species itself" (10). In his Conclusion to *Animal Rites*, "Postmodern Ethics, the Question of the Animal, and the Imperatives of Posthumanist Theory," Wolfe notes that the goal of such studies is not a new science of ethics: "As Derrida reminds us—and here he would be joined by every poststructuralist theorist I can think of—there can be no 'science' of ethics, no 'calculation' of the subject whose ethical conduct is determined in a linear way by scientific discoveries about animals (or anything else)" (190).

The kind of zoontological studies that interest Cary Wolfe are relatively recent and still very much underway (see for example, Scholtmeijer). During this interim, he remarks in his Conclusion to *Animal Rites*, he is "happy, practically speaking, to support the Great Ape Project (or the revision and upgrading of the United States Animal Welfare Act, or any number of other similar initiatives)," but while offering his support "only in abeyance, as it were, only in recognition of the underlying fact that the operative theories and procedures we now have for articulating the social and legal relation between ethics and action are inadequate—and here is the full posthumanist force of the question of the animal in this connection—inadequate for thinking about the ethics of *the question of the human as well as the nonhuman animal*" (192).

Zoographies

In *The Animal That Therefore I Am*, Derrida notes that for the Greeks, "zoography" referred not just to the painting of animals, but to portraiture of the "living in general" (12). Perhaps this sense of the word informs Matthew Calarco's choice of title for his 2008 book, *Zoographies: The Question of the Animal from Heidegger to Derrida*. For one of Calarco's main concerns in this book is the anthropocentrism that underlies liberal humanism and that leads to the ontological and ethical distinction of the "zoo" (*zōion*, animal) from the "human" self or subject. In search of new posthumanist understandings of animality that can effectively displace the human/animal dichotomy, *Zoographies* contributes significantly to critical animal studies, which Calarco describes as involving "a wide range of disciplines within the humanities, social sciences, and biological and cognitive sciences" (2), a field in which, in recent years, "traditional human-animal distinctions, which posit a radical discontinuity between animals and human beings, have been relentlessly attacked from multiple theoretical, political, and disciplinary perspectives" (3). As does Cary Wolfe, Calarco challenges the essentialism of Western concepts of subjectivity, or of what he calls its identity politics, with contemporary rights discourse and politics as an example of this: "Many animal rights theorists and activists see themselves as uncovering some sort of fundamental identity (for example, sentience or subjectivity) shared by all animals (or, rather, the animals they believe worthy of ethical and political standing) in order to represent that identity in the political and legal arena" (7). As well, he concurs with Wolfe on at least three other points: (1) what Derrida refers to as "the question of the animal" opens onto a large and rich set of issues that touch broadly on "the limits of the human;" (2) the "question of the animal" may well lead to the conclusion that none of our extant discourses is adequate for thinking and describing animal life and rethinking human-animal relations; and (3) animal studies is therefore, and necessarily, interdisciplinary, seeking out all available resources to aid in the task of thinking through "the animal question" (5–6). "There is no doubt that we need to think unheard-of thoughts about animals," Calarco writes, "that we need new languages, new artworks, new histories, even new sciences and philosophies. The field of animal studies is interdisciplinary precisely for this reason: it is seeking out every available resource to aid in the task of working through the question of the animal" (6).

Convinced that continental philosophy can contribute to this task, Calarco undertakes his study of by way of engaging the work of Martin Heidegger, Emmanuel Levinas, Giorgio Agamben, and Jacques Derrida. He proceeds with the thesis that "*the human-animal distinction can no longer and ought no longer to be maintained*" (3), assessing the writing of his four selected philosophers on animality and the human-animal distinction in relation to this thesis, interrogating critically how each philosopher establishes or displaces the distinction, and targeting the anthropocentrism that he thinks persists in much of today's animal studies discourse—as well as in the continental tradition he examines. In his argument, anthropocentrism should be the critical target of progressive thought and politics today (10), and, despite the rich resources it offers to animal studies, continental philosophy betrays its own "implicit anthropocentrism" (13). He takes it as his point of departure that, in order to move beyond the prevailing "metaphysics of subjectivity," it will be necessary to take seriously the "presubjective" realm, and to refigure the "presubjective" in radically nonanthropocentric ("postmetaphysical") terms.

Calarco's reading of Heidegger demonstrates what many critics neglect—that in his early writings, contending with the project of developing a fundamental ontology that would reground and reorient the human and biological sciences, as well as the university as a whole (31), Heidegger "does not take it as *philosophically* evident that there is a straightforward distinction to be drawn between human being and animal, or between living beings and nonliving beings" (21). Rather, Heidegger raises for serious discussion the question of whether the human-animal distinction can or should be drawn (23). For Calarco, this early work is radical and essential for Heidegger's subsequent work on "the animal question," as is a critical examination of where and how Heidegger later develops a "hyperhumanisim" of sorts (49), and an anthropocentrism that is far more complex than often recognized.[4] Calarco finds great promise in the work of Levinas, although it, too, particularly the discourse on the "face," betrays a stubborn anthropocentrism that many contemporary critics have faulted. Although Calarco deals with this discourse on the "face" and its contradictions, he gives novel importance to the thinking of ethics as "risk," the kind of "fine risk" that Levinas speaks of in *Otherwise Than Being: Or Beyond Essence*—the risk involved in focusing on animals even in an open-ended way, when "[t]here are no guarantees that we have gotten things right," or that any particular approach "will in fact have the kind of transformative effect we might desire" (77). In Calarco's chapter on Agamben, the difficulties of attempting a nonanthropocentric philosophy become evident again, for example, through Agamben's writings on the "anthropological machine," which focus "entirely and exclusively on the effects of the anthropological machine *on human beings* and never explore the impact the machine has on various forms of animal life" (102).

Finally, Derrida's work "serves both to further and limit the critique of anthropocentrism" Calarco advances throughout his book (14): it offers some of the most provocative and promising thought currently available to critical animal studies, but at the same time, Calarco says, Derrida, too, succumbs to anthropocentrism in that he does not rid philosophical thinking of the human-animal distinction. Indeed, Calarco takes "Derrida's insistence on maintaining the human-animal distinction to be one of the most dogmatic and puzzling moments in all of his writing" (145). This is a curious charge to level against a thinker for whom (given the "trace" and the "presubjective" that Calarco's discussion here all-but ignores), differences are primordial and cannot be reduced to some overarching sameness. This is the thrust of Derrida's remarks in the first passage Calarco cites from *The Animal That Therefore I Am* in support of his criticism of Derrida, a passage in which Derrida writes that throughout his work he has sought "to give, tirelessly, of my attention to difference, to differences, to heterogeneities and abyssal ruptures as against the homogeneous and the continuous. I have never thus believed in some homogeneous continuity between what calls *itself* man and what *he* calls the animal" (Derrida *Animal*, 30; qtd. in *Zoographies*, 145). I take Derrida's point here to be consistent with his overall critique of the Western tradition's assimilating of differences into would-be unities (differences between animals and animals covered over by "the animal" reification; differences between humans and humans subsumed under the metaphysical "subject;" and differences between animals and humans, covered over when "like us" standards are put in place). And it seems entirely out of keeping with Derrida's thought to suggest that his attentiveness to differences, including the multiplicity of differences that characterizes all of the living, is somehow equivalent to the reinstatement of a binary human/animal opposition.

Calarco's charge is particularly puzzling in light of his profoundly insightful reading of the scene with which *The Animal That Therefore I Am* opens, where Derrida relates the story of being caught naked in the gaze of his cat. "Derrida *does not know* who this cat is at the moment of the gaze any more than he knows who he is," Calarco writes (125). What the look of this animal gives Derrida (and hopefully his readers) to see is the other as absolute other, wholly other. Thus, as Calarco suggests, when Derrida says several times that the cat he is referring to is not a figurative cat, but a "real cat," his cat, his little, female cat, he is trying to gesture toward "something for which existing modes of thought are not particularly well equipped: the thought of this particular 'cat' as an absolutely unique and irreplaceable entity" (124). This particular cat is, finally, *unknowable*: "In insisting on the unsubstitutable singularity of the cat, then," Calarco writes, "Derrida is contesting the possibility of fully reducing this particular cat to an object of knowledge, whether philosophical or otherwise. Derrida *does not know* who this cat is at the moment of the gaze any more than he knows who he is. His encounter with the cat takes place in a *contretemps*, in a time out of joint, *prior to* and outside of knowledge and identification (125, emphasis mine). The "I," the subject who constitutes same/different, human/animal binaries, emerges *after* such "presubjective" exposure to the other animal—and to the abyssal difference of the other animal that can never be grasped, conceptually framed, or defined as the same, as "just like us." Thus, again as Calarco notes, for Derrida, "I am inasmuch as I am *after* the animal" (Derrida *Animal*, 10; qtd. in *Zoographies*, 125).

Zootobiography, Zoöpolis, Zoophilia, Zooanthropy

What Matthew Calarco calls the "scene of nonknowing" (125) with which *The Animal That Therefore I Am* opens is the space in which the autobiographical question, "Who am I?" is asked. For Derrida, what the look of the animal gives us to see is the impossibility of ever answering this question fully, of ever knowing the other animal, or of ever grasping the other in me so as to definitively answer the "Who am I?" question. Zootobiography, we might say, is the kind of autobiography that is written out of this nonknowing. It's the kind of autobiography we might expect from an animal, such as Derrida insists that he is in *The Animal That Therefore I Am*, a book that, as we have noted, collects the papers he presented at a conference on his work titled "The Autobiographical Animal." Zootobiography, then, does not appeal to the same framework that grounds the confessional discourse called autobiography and its self-referential, auto-affective self or subject, "the autobiographical and autodeictic relation to the self as 'I'" (Derrida *Animal*, 34). Perhaps it takes its genesis from some posthumanist version of the human animal, or as Derrida suggests at one point in *The Animal That Therefore I Am*, from the serpent of Genesis, who, when he says "I," is involved both in show and in disguise (65). For Derrida, the "I" presents (*graphs*) itself similarly, as always both an "I" and an other.

Zootobiography, not simply word play, raises some of today's most challenging philosophical questions, among them that of how to write and think ethics from other than the position of an authorial subject. Rather than have ethics originate with the spectating I/eye, Derrida suggests that we are first seen and called to respond to the look, appeal, of the animal other; that we write ethics as addressees of animals, rather than as their addressors. Perhaps it is as an addressee of animals that Jennifer Wolch, whose work we mention in an earlier chapter, envisions the twenty-first century "zoöpolis," a nonspeciesist urban center in which "wild" animals and humans coexist (see chapter 3), an urban space that models a nondiscriminatory ethics and practice. When, in "Witnessing the Animal Movement," their Introduction to *Animal Geographies*, Jennifer Wolch and Jody Emel suggest that a "zoophile" spirit infuses our contemporary culture, they are referring to a public sentiment that is both widespread and well-informed, and that leads increasing numbers of people to conclude that simply shooting or eliminating wild animals whose habitat has been displaced or destroyed by urban sprawl is no solution to the problems that humans themselves have created (10). Rather than denoting a kind of psychosis—the "*zoophil-psychosis* (literally 'love of animal psychosis')" that, Tom Regan points out in *Defending Animal Rights*, late nineteenth-century American neurologist Charles Loomis Dana diagnosed in those who felt deep empathy for animal suffering (Regan 133)—today's zoophile spirit may be our hope for tomorrow.

Despite the harrowing portraits of animal suffering that animal studies must needs confront, we can allow, too, that zoophila is in the air, that it is spreading through the disciplines and prompting sustained efforts by activists and academics to transform centuries of human domination of animals. It just might be that Cary Wolfe is on the mark when he suggests in *Animal Rites* that a hundred years from now "we will look back on our current mechanized and systematized practices of factory farming, product testing, and much else that undeniably involves animal exploitation and suffering" with the same kind of "horror and disbelief with which we now regard slavery or the genocide of the Second World War" (190). If that is the case, the various "zoo" studies surveyed in this chapter will have contributed to the change, these and many other efforts currently underway. The work selected for this chapter comes out of posthumanist critique and continental philosophy, from which just a few more writers might be mentioned in closing, the first being the hybrid philosopher (philosopher–actor–artist) H. Peter Steeves, whose work could well be given the label "zooanthropy" ("man-as-animalness"), rid of its heritage as some sort of delusion through which "a person believes himself to be an animal" (see Partridge 969).

The term *feral* is usually reserved for animals that cross the civilized/wild boundary, feral cats, for instance, that live both inside the outside of the domestic sphere (see for example, Griffiths, Poulter, and Sibley, "Feral Cats in the City"). In a number of instances, the term has also been applied similarly to children who belong both to wildness and to civilization—children raised by and as animals, and then rescued for domestication (Linnaeus in his 1735 *System of Nature* classified *Homo ferus* as a species). More than fifty cases of feral children are on record, Steeves notes in "Illicit Crossings," one chapter of his book, *The Things Themselves*, "human children raised in the wild by everything from bears and leopards to monkeys and birds" (19). The suffering and indignities that have been inflicted on these feral children is embarrassing, Steeves writes:

> The stories run from simple beatings and whippings (all in the name of "reinforcement training"), to the extreme cases such as the gazelle-boy, a human male raised by a family of gazelles, who, upon being captured, proved to possess the unnerving ability to leap great distances—jumping, nearly flying, through the air in the manner of his adoptive parents. His human benefactors, unable to persuade him to refrain from such activity and anxious to see him assimilated into human culture, considered their options and chose to cut the tendons in his legs, thereby inducing less gazelle-like behavior. (19)

Studies of such stories—each one of them different, Steeves points out (19)—belong to the task of uncovering traditional concepts of animality, of the human's elevation over "brute" life, and not the least, of the centrality of "speech" to demarcations of the human/animal binary.[5]

In "Lost Dog," which also appears in *The Things Themselves*, Steeves writes from Venezuela, where the city streets are filled with homeless dogs—not lost dogs, as he first took them to be, but dogs who "live in the city, in neighborhoods, and are watched over by many people" (50). These canine urban dwellers make him think of the "lost dog" story that Emmanuel Levinas tells in his essay, "The Name of a Dog, or Natural Rights," the story of "Bobby," so-named by the prisoners in Nazi camp 1492, whom the wandering dog visited for a few weeks before the guards chased him away. The dog would appear in the morning and be waiting for the prisoners when they returned from work, "jumping up and down and barking in delight," as Levinas himself puts it. While the Nazis treated Levinas and his fellow prisoners as less than human, for the dog "there was no doubt that we were men" (153). The dog showed the humans something like "respect," but for Levinas, as Steeves points out, since the animal nonetheless lacked "the brain needed to universalize maxims and drives" (Levinas 153), he remained an inferior being. "Indeed, Levinas has nothing respectful to say about animals in the short essay," Steeves maintains. "Bobby, we learn, has neither ethics nor *logos*. He is animal and therefore subhuman. He is (truly) what the Nazis were trying to make (falsely) their prisoners: 'a gang of apes,' 'no longer part of the world,' 'chatterers of monkey talk'— 'signifiers without a signified'" (50). Levinas, who wanted to deconstruct traditional binaries, ends up this essay "so often reconfirming their duality" (50).

Another of Steeves' texts, zooanthropic in its very title, "Rachel Rosenthal Is an Animal," is an essay (or is it a script?) in which, by approaching Rachel Rosenthal's performance art and painting (as well as the work of Plato, Shakespeare, Kafka, Robert Rauschenberg, Deleuze and Guattari, and others), Steeves "attempt[s] to understand the intricate relationships among the artist, the animal, the audience, and the notion of performance itself" (1). In the Introduction to his edited volume *Animal Others: On Ethics, Ontology, and Animal Life*, Steeves suggests that, although the book consists of a collection of essays, it really constitutes "a continuous text" in which, from the perspective of continental philosophy, the status of nonhuman animals is addressed under three main themes: "the limits of the body; animal ethics; and the relationship between the animal and philosophy—between, essentially, the animal and thought" (7). Along with Heidegger, Levinas, and Derrida, the philosopher whose work is engaged most prominently in this anthology is Maurice Merleau-Ponty, his thinking of the human-animal separation, subjectivity, nature, metaphysics, and alterity. One example of this engagement is Elizabeth Behnke's essay, "From Merleau-Ponty's Concept of Nature to an Inter-species Practice of Peace." To borrow the summary Steeves offers in his Introduction, Behnke's "phenomenology of living-with cats" leads her in this essay "to a rich reading of Merleau-Ponty; and her application of theory to the real-life case of a (potential) cat fight offers new directions for understanding what it is to experience a world together with animals, as well as the moral call to live appropriately in that world" (5).

We mention Alphonso Lingis in our discussion of Cary Wolfe's *Zoon-tologies*. Lingis is without doubt a zooanthropic thinker, who, in "Bestial-ity," his contribution to Steeves' *Animal Others* anthology, pondering the symbiosis of life with life, writes: "The form and substance of our bodies are not clay shaped by Jehovah and then driven by his breath; they are the coral reefs full of polyps, sponges, gorgonians, and free-swimming macrophages continually stirred by monsoon climates or moist air, blood, and biles" (39). The passage recalls another written by Lingis, this one taken from "Inner Space," an essay in which he perceives an awareness of such symbiosis in the work of sculptor Antony Gormley. For decades, Lingis points out, Gormley has been making casts of his own immobile body—standing, hands at his sides and feet together, lying on the floor, or crouched in a fetal position—sculptures that portray, not the dignity and authority of the heroic statues of humans that populate our public buildings and parks, but rather vulnerability (what Derrida calls passiv-ity), and that are not, then, at home in temples or museums, but outside, in natural environments, with animals. In the following passage from this essay, having much to do with the symbiosis of human and nonhuman life, and with autobiography as zootobiography, Lingis describes one of Gormley's works:

> *Another Place* consists of a hundred sculptures of iron that had filled seventeen moulds of Gormley's body held in an immobile position and differing only slightly, with individual traits of his features and his body effaced. They were first set up in 1997 over an area of 2.5 square kilometers in mud flats outside Cuxhaven, Germany, which was one of the major ports for emigration to America in the middle of the last [nineteenth] century. The fig-ures look outwards toward the horizon. Daily the tide submerges them; then they emerge again. Seaweed grows about them and

mussels attach themselves to them. Visitors are invited to them, to wander from one to another and to look with them toward the horizon of better places. We, with the sculptures, stand on this shore and are submerged and emerge from the mud of our life. ("Inner," 41)

7

FEMINIST CONTRIBUTIONS TO CRITICAL ANIMAL STUDIES

Ian Wilmut and Roger Highfield open their 2007 book, *After Dolly: The Promise and Perils of Cloning*, with the "tortuous tale" of the first mammal to be cloned from an adult cell by Wilmut himself and his colleagues. Born on July 5, 1996, the cloned animal, a sheep, "was (almost) genetically identical to a cell taken from a six-year-old sheep, the nucleus from which had been transplanted into an egg cell from a second sheep, and then inserted into the uterus of a third sheep, and then a fourth, to develop." Wilmut and Highfield go on to state that: "It was because the process started with a mammary cell from an old ewe that she was called Dolly—our affectionate tribute to the buxom American singer Dolly Parton" (13). Some feminists might not read this story as indicative of an "affectionate tribute," so much as of the sexism that still pervades our culture, including, perhaps especially, scientific laboratory culture. Moreover, as Sarah Franklin suggests in "Dolly's Body: Gender, Genetics, and the New Genetic Capital," the creation of Dolly takes sexism to a whole new level by linking it to a new kind of "breedwealth" that removes genetic capital from the animal herself, and shows how reproduction and genealogy can be owned, marketed, and sold (352–353). With nuclear transfer cloning, it becomes possible "for any animal, male or female, wild or domesticated, or even extinct, to become a perpetual germ-line repository, a pure gene bank," Franklin writes. "A single animal can be cloned to produce an entire herd of identical animals," not only compressing genealogical time, but also affording "total nuclear genetic purity, in perpetuity, and under patent" (353). Franklin argues that cloning raises important questions about maternity, paternity, gender, and sex—and lest we surmise that the process represents the triumph of maternity over paternity, she suggests that with cloning, "paternity has not so much been displaced as dispersed, into acts of scientific creation and principles of legal ownership. It may be the stud has vanished, but there are other father figures" (354).

Across the range of animality and animal welfare issues we have taken up thus far in this book, feminist perspectives have proven essential, and no doubt will continue to do so as the issues grow in number and complexity. However, feminism itself is as diverse a field as critical animal studies: interdisciplinary, marked by fluidity and by multiple theories and practices, so much so that attempts to classify the field are notoriously difficult. A number of typologies have been suggested nonetheless, for example the ones put forward by Josephine Donovan and Carol J. Adams in their Introduction to *The Feminist Care Tradition in Animal Ethics*, where their own approach, "the feminist care tradition," is considered alongside "postmodernist," "ecofeminist," and "Aristotelian" feminisms.[1] Since Carol Adams also counts herself as an "ecofeminist" (see Adams "Rare"), and is referred to in the Donovan and Adams Introduction as representing the "radical feminist tradition" (9), typologies such as theirs clearly allow for overlapping. Moreover, typologies may well label writers who have not themselves adopted the labels in question, as when Donovan and Adams identify James M. Coetzee and Jacques Derrida as two of "the most influential" proponents of the feminist ethics of care approach (14). Without attempting to overview either the vast field of feminist studies or the various typologies into which the field has been analyzed, this chapter foregrounds some of the ways in which feminist approaches to animal ethics go beyond, or offer alternatives to, both the utilitarian (interests) and deontological (rights) traditions that still prevail in animal studies.

Sexual/Textual Politics

Contemporary French feminist writers Hélène Cixous and Catherine Clément open their well-known essay "Sorties" by listing some of the binary oppositions that structure the Western metaphysical and patriarchal tradition—activity/passivity, sun/moon, culture/nature, day/night, father/mother, head/heart, intelligible/palpable, *logos*/pathos—all of these oppositions, they say, resting on this one: man/woman. "Always the same metaphor," they write: "we follow it, it carries us, beneath all its figures, wherever discourse is organized. If we read or speak, the same thread or double braid is leading us throughout literature, philosophy, criticism, centuries of representation and reflection" (63–64). And as many contributors to animal studies have pointed out, certainly feminists among them, the man/woman and man/animal binaries are inseparable in Western thought. Thus, Lynda Birke and Luciana Parisi contend in their essay, "Animals, Becoming," that dualism, the predominant Western tendency to think in terms of either/or oppositions, is what brings feminist studies and animal studies together in a common concern: "Such dualisms pervade our thought, and all are hierarchical. So, both feminism and the causes of animals must share a concern with the ways that the Other becomes subordinate" (55).

The problem of moving beyond anthropocentrism and the centuries-old human/animal hierarchy has preoccupied feminists for several decades now, so both feminist studies and critical animal studies involve questions of how to inherit tradition. In their essay, for example, Birke and Parisi trace the dualism of the Western tradition back to Plato, Aristotle, Descartes, Kant, Darwin, and others, thinkers whose work is now being widely reread for contributing to the human/animal dichotomy. In both feminism and critical animal studies, we find agreement as to the violence and repression that dualistic thinking entails, such that, as Cixous and Clément write, "the movement whereby each opposition is set up to make sense is the movement through which the couple is destroyed. A universal battlefield. Each time, a war is let loose. Death is always at work" (64). In the late 1960s and through the 1980s, French feminist criticism—linked in one way or other to psychoanalytic theory and poststructuralist studies—undertook an extended evaluation and critique of logocentric binaries as embedded in language and writing, both of the latter understood not simply as transmitters of ideas, but as constitutive of them and of subject identity. In other words, French feminist writers of that period concerned themselves particularly, and in many different ways, with the relationship between "textuality" (broadly understood to include all signifying systems) and the structuring of a logocentric (rational and masculine, speech-centered) self-identity. For a crucial two decades or more, these concerns distinguished French feminists from their American feminist counterparts, a distinction Toril

Moi caught in the title of her 1985 book, *Sexual/Textual Politics*. Moi's title refers to Anglo-American feminists as, on the one hand, engaged in an activist identity politics that often advocated strongly for women's rights, and that was "mostly indifferent or even hostile" (70) toward the theoretical emphasis of French feminists; and French feminists as, on the other, engaged in what Moi calls a "textual politics," a politics based on the conviction that identity is not biologically given, and that a radical challenge to patriarchy required a focus on the "subject," its "textual" formation and its potential for transformation.

There is no homogeneity across the work of those who have been labeled "French feminists," including, among others: Cixous, Clément, Marguerite Duras, Shoshana Felman, Lucette Finas, Luce Irigaray, Sarah Kofman, Julia Kristeva, and Annie LeClerc. To a large extent, their work predates the current animal studies movement, and yet, given the historical associations between "woman" and "animal," their analyses of the West's self/other, mind/body, man/woman binaries provide valuable resources for critical animal studies. One example is the work of Bulgarian-born French feminist and psychoanalyst Julia Kristeva. In 1973, she published the first section of her state doctorate thesis, translated into English the following year as *Revolution in Poetic Language*, which outlines what she calls, in French, *sémanalyse*—that is, a method for approaching language and texts (including, we would have to say, ethics texts) not as products or tools for representing "reality," but as material systems through which "reality" is *produced*. Drawing on the work of Sigmund Freud and Jacques Lacan, *Revolution in Poetic Language* develops her notion of language as a dialectical struggle between two poles, the "semiotic" (unconscious) and the "symbolic" (propositional, representational) modalities. The important points are that, for Kristeva, first of all, language is not an instrument or a unitary thing, but a struggle between these two modalities; and, second, that Western culture, with its investment in the "rational" or representational pole of language, has consistently repressed and disallowed the unconscious, embodied dimensions of language and subject identity. Not incidentally, the same "Cartesian" culture has associated "woman" with "body," with the dark, underside, unconscious, nonrational, "animal" realm. "Woman" and "animal" belong together as the repressed of Western culture, not the least throughout Western modernity.

Kristeva draws out this link between "woman" and "animal" in *Powers of Horror: An Essay on Abjection*, giving particular attention to the repressed (embodied, "animal") dimension of language and self identity, and using the word *abjection* to refer to the subordination, rejection, or violent *expulsion* of one's own animality each time the human constitutes itself as a rational, integral "I." This powerfully repressive abjection is repeated over and over again in language, with each positing of subject identity as unified rationality, with each attempt to keep the other (animal, woman) outside of the self-same. "The abject confronts us," Kristeva writes, "with those fragile states where man strays on the territories of the *animal*" (12), the threatening world of animals or animalism that, in order to become a unitary "I," one must violently reject. In *Tales of Love*, she deals with the opposite, idealizing, pole of language and narcissistic subject formation, exploring historical myths of love in order to probe the significance of idealizing fantasies (such as the fantasy figure of the self) for the process of constituting identity and meaning in binary either/or (man/animal, man/woman, same/different) terms. Kristeva's analysis, informed by psychoanalytic theory, suggests something of the significance of Christianity and its discourse on love for Western culture's fantasy of an ideal, homologous, thus disembodied, "I." For in Christianity, she says reading Paul, the faithful are enjoined to identify with the supreme ideal (God, the risen Christ), and to do so by allowing "the lustful body, the erotic body" to die, "in order to recover, through resurrection, the body in its integrity but completely invested in the ideal" (142). The bond the tradition establishes between the divinity and its believers becomes "an *identification*," she suggests, and, from a psychoanalytic perspective, there is no such identification without the sacrifice of a body, an animal body, "one's own body" (143).

Embracing quite a different kind of feminism than that of Kristeva, American (Minnesota)-born Kate Millett published *Sexual Politics* in 1969, becoming, as a result of the book, a reluctant figurehead for the "new" ("second wave") Anglo-American women's movement. According to Millett, winning the vote did not free women, any more than blacks, from the domination of white men; she called for the resurgence of a feminist struggle to wrest political power from men. As Toril Moi notes, for some years after its publication, Millett's book went unchallenged by feminists in England and America, who thus, at least implicitly, supported its vehement dismissal of Freud and its hard-hitting thesis that feminism is all about sexual power-politics (26). Using Millett's book title, Moi characterizes Anglo-American feminism, from Millett through the 1970s and early 1980s, as a "sexual politics" rooted in traditional humanism, its faith in realist forms and its essentialist tendencies. What the work of such feminists did not grasp, she suggests (her examples here are Elaine Showalter and Marcia Holly), "is that the traditional humanism they represent[ed] is in effect part of patriarchal ideology. At its center is the seamlessly unified self—either individual or collective—which is commonly called Man" (8). As a struggle for political power, Anglo-American feminism was well disposed through the 1970s and early 1980s to adopt "rights theory" and to advocate for women's rights in all cultural and institutional situations.

By the mid-1980s, however, rights theory itself faced challenges from within the women's movement, and it became impossible to say that Anglo-American feminism translates simply into a struggle for rights—for women or, by extension, for animals, whose cause feminists had taken up by this time. On the contrary, as Tom Regan has it, his animal rights argument has been "indicted" by Anglo-American feminists, on the grounds that the idea of individual rights, itself patriarchal, "encapsulates male bias" (Regan *Defending*, 53) and embeds the "dualistic, hierarchal rankings men tend to make" (55). Philosopher Angus Taylor writes in his book, *Animals and Ethics*, with reference both to Regan's rights theory and Singer's utilitarianism, that feminists find these approaches "inadequate, limited by their association with dominant philosophical traditions" (78). In the view of philosopher and women's studies professor Elizabeth Anderson, writing in "Animal Rights and the Values of Nonhuman Life," neither Singer's utilitarianism nor Regan's theory of rights, "has successfully generated a valid principle of action that does justice to all the values at stake" in the context of human and nonhuman animal relations (279). For her, both approaches are "simplistic" in attempting to derive principles of justice from a single criterion, "the possession of certain valuable capacities" (280), rather than opening rights and interests claims, as they should be opened, to "a plurality of qualitatively different values" (290).

Perhaps, as its title suggests, Carol Adams' *The Sexual Politics of Meat*, belongs to a sexual politics—to the humanist tradition and its essentialist model of sexual difference—for the book does argue that meat-eating is male, a symbol of virility and masculine power, and that vegetarianism is a (female) feminist strategy, a reverse "sexual politics." M. E. Grenander, in a review of *The Sexual Politics of Meat*, refers to Adams as "a powerful champion of animal rights" (335), and Adams herself, in a later reflection on *The Sexual Politics of Meat*, suggests that at the time she wrote the book, "[r]ights, it seemed to me, at least said, minimally, 'do not touch.'

And there are many times when that is exactly what I feel about the exploitation of animals. I want to intervene and proclaim this" (Donovan and Adams, *Feminist*, 199). Along the way, however, and in the course of authoring and editing numerous books—among them *The Pornography of Meat, Neither Man Nor Beast: Feminism and the Defense of Animals*, and *Animals and Women: Feminist Theoretical Perspectives*, edited with Josephine Donovan—Adams began to study herself and animal issues more closely, coming to the conclusion that "the male ideal of the autonomous individual, on which rights theory is based, is fraudulent" (199). Safe to say, then, taking Adams as a case in point, the sexual/textual distinction that, at one time, differentiated French feminists from their Anglo-American counterparts, no longer remains neatly intact. Or, to put the point in another way: both "French" and "Anglo-American" feminisms have moved beyond subject-centered traditions of ethics, offering significant historical and theoretical critiques of anthropocentrism, speciesism, and the hierarchical (mind/body, man/woman, man/animal) dualism this subject inevitably reinstates.

Ethics of Care

Carol Adams elaborates her more recent critique of rights theory in favor of a feminist "ethics of care" in the above-mentioned anthology she edited with Josephine Donovan, *The Feminist Care Tradition in Animal Ethics*. The book—a sequel to the 1996 collection edited by Donovan and Adams, *Beyond Animal Rights: A Feminist Caring Ethic for the Treatment of Animals*—includes many of the chapters that appeared in the 1996 volume along with chapters that respond to it, and/or outline "new theoretical developments" (1) in the feminist animal care approach. The book provides a good overview of the features that set the feminist care tradition apart, features that we will briefly consider here—some suggestive of a movement beyond the tenets of humanism, some retaining humanist concepts, albeit differently, feminist care theorists maintain.

For example, a dual, but not dualist, account of sexual difference is evident in feminist ethics of care theory, a theory that, according to Donovan and Adams in their Introduction to the 2007 book, traces its origins to Carol Gilligan's *In a Different Voice*. Gilligan's book, they explain, "identifies a women's 'conception of morality' that is 'concerned with the activity of care . . . responsibility and relationships,' as opposed to a man's 'conception of morality as fairness,' which is more concerned with 'rights and rules'" (Donovan and Adams *Feminist*, 1–2). Tom Regan comes under attack here. Donovan argues that his rights theory "privileges those with complex awareness over those without," and "depends on a notion of complex consciousness that is not far removed from rational thought, thus, in effect, reinvoking the rationality criterion" of the Cartesian-Kantian tradition (62). Stemming from natural rights theory, Regan's animal ethics privileges rationalism and individualism, and Donovan notes, it excludes "sentiment" from "'serious' intellectual inquiry" (62).

Rejecting rights, and "abstract, rule-based principles," the feminist care tradition favors a "situational, contextual ethics" (2). Still, this is not the "situation ethics" advocated by Peter Singer, whose approach also faces significant critique in the 2007 Donovan-Adams volume. Writing in that volume, Deborah Slicer refers to the "Singer-Regan approach," which she characterizes as essentialist and based on a self-centric moral standard of sameness: "Singer and Regan extend the moral community to include animals on the basis of sameness. They do not acknowledge, much less celebrate, differences between humans and other animals" (109). They rely on "general, prescriptive principles" that are largely indifferent to context, oversimplifying, and presented as if they were the only options available to ethics today (111–113). And not the least, in line with the "masculinist contempt for our emotions," both fail to account for love, friendship, feelings—a variety of affective responses—as appropriate and necessary to animal ethics (113–114).

In the 2007 Donovan and Adams volume, addressing what she calls "the war on compassion," Carol Adams traces human indifference to animal suffering to the "ability to objectify feelings, so they are placed outside the political realm" (33). In the feminist care tradition, on the contrary, compassion is theorized as inescapably political. *Attention* is the key word here, Donovan and Adams explain: "Attention to the individual suffering animal, but also—and this is a critical difference between an ethic-of-care and an 'animal welfare' approach—attention to the political and economic systems that are causing the suffering. A feminist ethic-of-care approach to animal ethics offers a *political* analysis" (3).[2] Certainly, political implications are evident in the "core principles" that Donovan and Adams identify as supporting the feminist care tradition. The first says that it is wrong to harm sentient creatures "unless overriding good will result *for them*," and that it is wrong to kill sentient creatures "unless in immediate self-defense or in defense of those for whom one is personally responsible" (4). The second principle states that, "humans have a moral obligation to care for those animals who, for whatever reason, are unable to adequately care for themselves, in accordance with their needs and wishes, as best the caregivers can ascertain them and within the limits of the caregivers' own capacities" (4). Third, "people have a moral duty to oppose and expose those who are contributing to animal abuse" (4). It follows from these principles that an anti-vegetarian position is incompatible with a feminist ethics of care, "because those who care about animals obviously do not destroy and eat them" (13). The tradition thus opposes farming, both traditional and industrial, and animal experimentation (unless of benefit to the individual animal involved). It morally requires both active care for animals, such as through shelters and rescue initiatives, and political activism to expose and oppose animal abuse.

Cyborg Science

Why has Western culture been so concerned, conceptually and morally, to separate humans from "other animals"? Biologist Lynda Birke addresses this question in *Feminism, Animals, and Science: The Naming of the Shrew*, a study that, focusing on "the *idea* of 'animal,'" and its link to work on feminism and science (6), analyses critically the human/animal separation that culture and science have so long maintained. In the course of this analysis, appealing to the argument that women and animals are "fellow sufferers" (135), Birke endorses a feminist ethics of care, sympathy, and compassion over against the rationalism of the individualist rights tradition and of reductionist, determinist, and increasingly commercialized science. Hers is one of a number of contributions to critical animal studies by feminist scientists whose work interrogates the meaning of science as it relates to gender, body, and "animal" (see, for example, Guerrini; Harding; Keller), work that inevitably confounds the hierarchical human/animal (man/animal) dichotomy.

Donna Haraway's work, as a case in point, brings feminist theory and animal studies together with science, technology, and biomedical studies in ways that displace the mind/body, man/woman, man/animal binaries in favor of a "cyborg" vision—an analysis that demonstrates the constructed nature, and scientific inadequacy, of history's masculine rhetoric of rationality, technology, invasion, and war in favor of ideas of caring and of fruitful biological interconnections. If, as Lynda Birke suggests (*Feminism*, 144), Haraway's work is "playful" (a term often associated with *postmodernism*), its theoretical and political import are nonetheless significant. For example, her 1991 book, *Simians, Cyborgs, and Women: The Reinvention of Nature*, interlacing her knowledge of social sciences theory (Karl Marx, Marxism, Frankfurt School Critical Theory, Fredric Jameson), feminist theory (Simone de Beauvoir, Judith Butler, Nancy Chodorow, Teresa de Lauretis, Audre Lorde, Catherine MacKinnon, Juliet Mitchell, Monique Wittig), the history of human-animal relations, technology, and the life sciences, is multidisciplinary and challenging—in the best sense of that word. The book introduces a number of feminist approaches to women's studies and animal studies, including approaches that Haraway herself has adopted at different stages of her career. Thus, the essays in the first part of the book come out of the 1970s, when Haraway was "a proper, US socialist-feminist, white, female, hominid biologist, who became an historian of science to write about modern Western accounts of monkeys, apes, and women"; while at the time of the last essays, she had "turned into a multiply marked cyborg feminist" (1). Granted, the cyborg feminist remains in many ways a Marxist-socialist feminist, but an always-explorative fluidity is evident and it testifies to one thesis of the book: identities, rather than unified and static, are constructed within contested discourses; they are constantly in flux, contradictory, partial, and strategic (see, for example, 155).

With Haraway, then, feminism approaches animal studies by first abandoning the humanist notion of the fixed and pregiven self. Through the pages of *Simians, Cyborgs, and Women*, in essays that chart her transition from socialist feminist to cyborg feminist, she examines the "breakup of versions of Euro-American feminist humanism in their devastating assumptions of master narratives deeply indebted to racism and colonialism," turning the book's concern to the possibilities of a cyborg feminism "that is perhaps more able to remain attuned to specific historical and political positionings and permanent partialities without abandoning the search of potent connections" (1). Both in her critique of the humanist subject and in her theorizing of a cyborg feminism, Haraway makes a distinct contribution to critical animal studies, one that is evident not only in the 1991 book, but also in a number of her other publications, including *Primate Visions: Gender, Race, and Nature in the World of Modern Science*; *The Companion Species Manifesto: Dogs, People, and Significant Otherness*; and *When Species Meet*.

For Haraway, not only identities but also bodies are made, historically specific constructions: organisms do not preexist, she argues in chapter 10 of *Simians, Cyborgs, and Women* "The Biopolitics of Postmodern Bodies: Constitutions of Self in Immune System Discourse," and "even the most reliable Western individuated bodies, the mice and men of a well-equipped laboratory, neither stop nor start at the skin" (215). Since about 1950, in biology and medicine as well as in other political and cultural contexts, she explains, the naturalized body of humanism's rational self has given way to the body as a "coded text," or "semiotic system, a complex meaning-producing field" (211), accounts of which involve "recognition and misrecognition, coding errors, the body's reading practices (for example, frameshift mutations), and billion-dollar projects to sequence the human genome to be published and stored in a national genetic 'library'" (211). In light of such current understandings of the biomedical body, immune system discourse, "that is, the central biomedical discourse on recognition/misrecognition, has become a high-stakes practice in many senses" (211).

From a classical medical perspective, the immune system is characterized by the development of antibodies that are produced in response to foreign invasion of the boundaries of the self, for example, by a threatening virus or some other invasive pathogen. In these terms, a well-functioning immune system distinguishes what is self from what is not self, and neutralizes or eliminates that which is foreign or other. Corresponding to this understanding of the immune system is the semantics of defense and invasion, aggressive, military Star Wars–type imagery—and a notion of the self as a fully defended and bounded territory (224). But as Haraway maintains in this chapter, "what does not count as self" is no longer clear, and it is precisely the boundaries between self and other that, now recognized as generated rather than given, are constantly in negotiation and flux. Thus, some current research suggests conceiving of the immune system as a "potent mediator, rather than a master control system or hyper-armed defence department," perhaps even as a "sensory organ" (252n6). "No wonder auto-immune disease carries such awful significance, marked from the first suspicion of its existence in 1901 by Morgenroth and Ehrlich's term, *horror autotoxicus*" (222–223).

As Haraway also points out, immunology is not only a medical specialization, "a subject of research and clinical practice of the first importance" (205), but also a biopolitical and cultural trope through which all kinds of self and other boundaries are being posited and contested, "an elaborate icon for principal systems of symbolic and material 'difference' in late capitalism" (204). Metaphors of pregiven and intact individual units and borders under surveillance by a defense system poised to attack invaders no longer work in these discursive spheres either. For, where "[w]hat counts as a 'unit,' a one, is highly problematic, not a permanent given" (212), we can no longer always differentiate "outside" from "inside," and an attack on a perceived invader might then turn out to be a deadly autoimmune assault on that which actually lies "within." Based on Haraway's analysis here, as well as Jacques Derrida's writing on auto-immunity ("Autoimmunity," *Rogues*), we might say that the metaphysical concept of the self of Western tradition is autoimmune, an attempt to attack and cast out (abject) an invader or foreigner, "the animal," that actually inhabits the self from within. And as the contemporary environmental and ecological crisis indicates, the attack is proving deadly.

The late twentieth-century biomedical and cultural body, as Haraway portrays it, can be conceived either in lingering humanist terms—"master control principles, articulated within a rationalist paradigm of language and embodiment"—or "in terms of complex, structurally embedded semiosis with many 'generators of diversity' within a counter-rationalist (*not* irrationalist) or hermeneutic/situationist/constructivist discourse readily available within Western science and philosophy" (213). It is to the latter mode of thinking that her cyborg belongs. In chapter 8 of *Simians, Cyborgs and Women*, "A Cyborg Manifesto: Science, Technology, and Socialist-Feminism in the Late Twentieth Century," she articulates her cyborg vision in some detail, explaining that what she calls a *cyborg* is "a hybrid of organism and machine" (149), two realms that the Western tradition has held resolutely apart. For her, the cyborg is specifically a late twentieth-century development, emerging from a world in which "organic creatures" and "high technology" can no longer be bordered off one from the other, organic creatures having assumed the guise of "information systems, texts, and ergonomically controlled labouring, desiring, and reproducing systems" (1). Like the Canadian political philosopher George Grant (1918–1988) and the French philosopher Jacques Ellul (1912–1994), Haraway suggests that, by late modernity, technology has so thoroughly penetrated biological and social life that it has "become us," definitive of our self-understanding and ways of being in the world. But unlike Grant and Ellul, Haraway does not lament this development, something that should come as no surprise given the discussion above. "The cyborg incarnation is outside salvation history," she writes: it does not belong to narratives of a fall, apocalypse, or redemption. Yet ("Marx-ist," perhaps in this sense), cyborg thinking is utopian; indeed, under-standing cyborg logic, she says, may be the key to our survival. Chapter 8 of *Simians, Cyborgs, and Women* makes an argument "for the cyborg as a fiction mapping our social and bodily reality and as an imaginative resource suggesting some very fruitful couplings" (150). Her argument is based on the assumption that by the late twentieth century, "we are all chimeras, theorized and fabricated hybrids of machine and organism; in short, we are cyborgs" (150).

As a hybrid of organism and machine, a cyborg certainly cannot be conformed to the Cartesian mind/machine opposition. Indeed, rather than positing the body (human or animal) as *but* a machine, exterior to essential selfhood, an object that subjects control, Haraway confounds the Cartesian binary by suggesting that the bifurcating of mind from matter-machine is illusory. Arguing for "*pleasure* in the confusion of boundaries" such as these (150), she singles out three crucial boundary breakdowns in particular, each of which is relevant for critical animal studies. First: "By the late twentieth century in United States scientific culture, the bound-ary between human and animal is thoroughly breached" (151). Sec-ond, the boundary "between animal-human (organism) and machine" has collapsed: "Late twentieth-century machines have made thoroughly ambiguous the difference between natural and artificial, mind and body, self-developing and externally designed, and many other distinctions that we used to apply to organisms and machines" (152). Third, "the bound-ary between physical and non-physical is very imprecise for us," as, for example, modern machines become smaller, "quintessentially microelec-tronic devices," more invisible and more ubiquitous, many of them "all light and clean because they are nothing but signals, electromagnetic waves, a section of a spectrum" (153).

In the face of these rapid changes and disappearing boundaries, some might say that, "a cyborg world is about the final imposition of a grid of control on the planet, about the final abstraction embodied in a Star Wars apocalypse waged in the name of defence, about the final appro-priation of women's bodies in a masculinist orgy of war" (154). But from Haraway's perspective, "a cyborg world might be about lived social and bodily realities in which people are not afraid of their joint kinship with animals and machines, not afraid of permanently partial identities and contradictory standpoints" (154). The political struggle (for women's studies as for animal studies) is "to see from both perspectives at once because each reveals dominations and possibilities unimaginable from the other vantage point. Single vision produces worse illusions than double vision or many-headed monsters" (154).

Neurological-Ontological Contortions

Historian and philosopher of science Evelyn Fox Keller writes on the back cover of Australian neurologist Elizabeth A. Wilson's *Psychosomatic: Feminism and the Neurological Body* that: "It is quite a while since we have heard a voice as refreshing as that of Elizabeth A. Wilson. With boldness, wit, and extraordinary inventiveness, she shows us just how delimiting have been prevailing tendencies in science studies and feminist theory to marginalize, if not outright repudiate, the material, biological dimensions of human psychology." Keller suggests that Wilson's book "opens" the life sciences, especially neurology, "to dramatically new understandings." Although Wilson's study centers on delimiting tendencies in the life sciences, the "dramatically new understandings" her book offers pertain as well to critical theory (she is a critical theorist herself), the humanities, and the social sciences—particularly as to prevailing accounts of embodiment, mind-body relations, sexual difference, selfhood, and not the least, human-animal difference. A key question the book raises (one of several) has to do with *psyche*: who or what is the psyche embedded in the book's title, and how does she/it relate to critical animal studies?

For Wilson, to address psyche is, necessarily, to read Freud—whose work, we note earlier, is all but missing from critical animal studies. Wilson opens her book by returning to the early Freud, going back to the 1870s when, still a medical student at the University of Vienna, he began working in Ernst Brücke's physiology lab, dissecting the spinal ganglia of the sea lamprey, a primitive petromyzon fish, and demonstrating, as a startling result of this research, a continuity in the nerve cells of "higher" and "lower" animals. It is here, in Freud's neurological-anatomical studies, rather than in his later work on hysteria, that Wilson locates the beginnings of psychoanalysis—which means that, for one thing, Wilson traces the emergence of psychoanalysis to Freud's speculation concerning a *continuity* between human and animal life, a novel suggestion in its own right. More precisely, she traces the beginnings of psychoanalysis to "the spinal ganglia of the petromyzon" (1)—to an animal, let's say, although not all critical animal theorists would concede animal status to a primitive fish. But the nervous system of Freud's fish, she argues, communicates past the 1870s, "past the dissection table" (3), and even, we might add, into critical animal studies today.

Wilson's Introduction turns to feminist accounts of hysteria, particularly the conversion hysteria that Freud documented and theorized on his own, and in collaboration with Joseph Breuer. Working through a number of feminist texts, she notes that they tend to theorize "the body," especially the hysterical body, as primarily ideational (5). They shun the biological (7) and pass over neurophysiology (8). Wilson then considers "The Case of Fraulein Elisabeth von R.," focusing on the importance of biology, and of the biological body, in Freud's analysis of this case of conversion hysteria, where Fraulein Elisabeth's muscles became, as it were, "psychologically attuned" (9), and where her nerves, blood vessels, and skin facilitated her "psychological" illness (10). It is significant, for Wilson, that, in observing these features, Freud does not "take refuge in the central nervous system," that is, he does not look to "the brain or higher cortical surfaces" for an explanation of conversion (10). "It is as though he suspects that the psychic conflicts have been devolved to the lower body parts: here, psychic defense is more muscular than it is cerebral" (10). We might say then, that for Freud, as he puts it in the last note he wrote before his death, "Psyche is extended" (see Derrida's discussion of this extension in *On Touching* and Jean-Luc Nancy's consideration of psyche in *Corpus*). Psyche is not "the mind," located in the "higher" brain; Wilson suggests that psyche cannot be located through a high/low hierarchy. She/it is dispersed: if Fraulein Elisabeth's pains are all "in her head," Wilson writes, "then this entails a number of reciprocal ontological contortions: that her thigh is her head, that her mind is muscular, and the Freud's words are in the nature of her nervous system" (11). Moreover, Wilson suggests, Fraulein Elisabeth's psyche extends past her body, such that, "the nervous systems of animals might be useful for critically astute formulations of human bodies" (13). This is a proposition that she explores in the subsequent chapters of her book, which open to additional dislocations of the mind/body, man/animal hierarchy.

Animal Pedagogy

Kelly Oliver's *Animal Lessons: How They Teach Us to Be Human* opens with a reference to the October 2003 *Siegfried & Roy* show in Las Vegas during which Montecore, a white tiger who had performed in the show for years, attacked and bit Roy, dragging him off stage by the neck (1). Several pages later, Oliver states that her project in *Animal Lessons* is "to indicate various ways in which, despite philosophers' attempts to domesticate the animal, animality, and animals like Montecore, animals break free of their textual confines and bite back" (8). In making this case, Oliver challenges central assumptions about humans and animals that prevail in animal studies today, looking "to an animal ethics that disarticulates the ways in which the concepts of animal, human, and rights all are part of a philosophical tradition that trades on foreclosing the animal, animality, and animals" (22). In part, for her, the work of disarticulating or "deconstructing" tradition is guided by animals themselves—animals who, in escaping philosophical confines, have much to teach humans about themselves, and about alternate ways to think animality and animal ethics. Drawing comprehensively and critically from continental philosophy—including the work of Derrida, Heidegger, Merleau-Ponty, Agamben, Freud, Lacan, and Kristeva—Oliver contributes significantly in this book to the development of what she calls a "free-range ethics," one that tears down the fences constructed by "the self-centered, exclusionary, and domineering notions of individuality, identity, and sovereignty" (305), and that foregrounds relationships, sharing, sustainability, and responsibility.

For instance, in one chapter, "Sexual Difference, Animal Difference: Derrida's Sexy Silkworm," Oliver examines the Western philosophical tradition's construction of "the war between the sexes," its man/woman dualism that typically locates "woman," along with "animal," on the underside of the binary divide. The man/animal dualism reduces the multiplicity of living beings called *animal* to a single entity, opposing that to "man," just as the man/woman hierarchy reduces multiple sexualities to a two-term, either/or binary. Recognizing that in the Western tradition these two binaries (man/animal, man/woman) are intimately connected, Oliver wonders whether animals might teach us to imagine sexual difference beyond dualism: "if we consider various sexes, sexualities, and reproductive practices of animals, we might be able to reconsider the sexes, sexualities, and reproductive practices of humans beyond the tight-fitting binary of man/woman or homosexual/heterosexual. My thesis is that the binary oppositions man/animal and man/woman are so intimately linked that exploding the first has consequences for the second" (132).

Jacques Derrida's work on the connection between sexual difference and animal difference is a crucial resource for Oliver in this chapter,
where she considers Derrida's reading of Heidegger on the questions of
ontological and sexual difference, as well as Derrida's writing on *différance*
and his attempts to conceive of difference differently, in a nondual way.
As she points out, and with obvious relevance for critical animal studies,
what Derrida calls a "hyperbolic ethics," an ethics of difference, "cannot
be fixed into a set system of discernable characteristics. Instead, what
counts as different or distinct must remain an open question" (139). It
follows that "ethics cannot begin with the couple, pair, or binary. Neither
can it begin with one or three, other numbers favored by philosophers"
(139). The number three belongs to the triple movement of the Hegelian dialectic, the system Oliver analyses in her 1997 study, *Family Values*,
demonstrating how woman and animal belong together for Hegel as the
negation of man, as the "middle term" that, as Derrida puts it in *Glas*,
"disappears" in the final movement of the dialectic (Oliver *Family*, 34–46;
Animal, 133; Derrida *Glas*, 170a).

But the insightful surprise in this chapter comes with Oliver's appeal
to the "multitude of sexualities and sexual differences beyond two"
(147–148) that are found in the world of animals, among worms and
damselflies, for instance, who have something to teach us about thinking difference differently. Here, she refers not only to Derrida's writing
on silkworms as beyond the sexual binary, but also to a recent article on
ocean worms that have evolved at least eighteen different ways to reproduce (148). "In a world where there are millions of species of worms with
dozens of modes of reproducing and where damselflies can change sex
or reproduce their 'better half,' what happens to our traditional notions
of male and female, or binary sex?" (149). Oliver's point is not only to
demonstrate that all animals cannot be divided into the binary couple
male/female, but also to encourage her readers to imagine "the possibility of alternative sexes and sexualities," to open to alternatives beyond
"the limited and claustrophobic binary that reduces sex to a war between
two" (150). In this chapter, and throughout the book, animals are Oliver's
teachers. They help us to imagine an ethics based on differences that cannot be contained by a binary.

8

KILLING AND EATING ANIMALS

Perspectives from World Religions

In 1940, four teenage boys walking with a dog in southwestern France happened upon the opening to what proved to be a prehistoric cave. Scholars have since dated the Lascaux cave to the Magdalenian period of the Upper Paleolithic era, some 16,000 to 20,000 years ago. Housed within the cave, on the walls and ceilings of its rooms, are thousands of paintings and etchings of animals, including horses, stags, aurochs, ibexes, bison, bears, and felines, some of them, such as the seventeen-foot-long bull found in the chamber that has been named the Hall of the Bulls, breathtaking in their majesty. The Lascaux images belong to a hunter-gatherer stage of prehistory when animals were killed and eaten—the Magdalenian people lived amid large herds of wild horses, bison, and reindeer, and survived by hunting and fishing—yet archaeologists have suggested that, although these hunter-gatherer people killed and ate animals, they nonetheless held them in high esteem. For example, archaeologist and professor of early prehistory, Steven Mithen, referring to images in the Lascaux cave and in the even older cave of Chauvet-Pont-d'Arc—discovered in 1994, also in southern France, and dated to approximately 31,000 BCE—contends that although the meaning of these paintings may be beyond our powers of deduction, their quality is "astounding," and their import relates certainly to the rich mythological world of these prehistoric people. For him, the paintings reflect an aesthetic appreciation of the "beauty and grace" of animals, and along with that, a collapse of any sharp boundaries between humans and animals that may have existed during earlier evolutionary stages ("Hunter-Gatherer," 123; see also Bahn). Mithen refers in particular to a 30,000-year-old painting from the Chauvet cave that depicts the head and shoulders of a bison with the legs of a man, this as suggestive of collapsed human/animal boundaries (123). The bison/man painting also suggests that, "humans of 30,000 years ago were able to conceive of entities that broke the rules of nature, entities that could not exist in the physical world. In other words, they were engaging in the type of imagination that can be described as fantasy" (Mithen "Evolution," 48) The mythological imagination Mithen attributes to these prehistoric cultures recalls John Berger's words (from *Why Look at Animals?*) that, as "the first subject matter for painting" (5), animals likely entered the imagination of primeval people, not as meat or horn, but as messengers and promises (2).

On a similar note, anthropologist André Leroi-Gourhan points out that while Paleolithic cave art was seen until recently as "magic in nature," it now "appears to have been a more general figurative system, a real mythology" in which, for example, the images are "connected with one another by the link of a theme whose meaning escapes us but which is repeated again and again" (326–327). Does this art represent a real cosmogony? Leroi-Gourhan cannot answer the question definitively. Kimberley Patton asks: "Do these murals belong to the realm of religion?" She thinks that there "can be no doubt" but that they do, basing her opinion on the broad definition provided by Clifford Geertz and Evan Zuesse that religion comprises "systematic thought that orients human existential experience to metaphysical powers through external, culturally accepted forms" (Patton "Caught," 30). The latter definition is but one of many that continue to be debated within the field of the academic study of religion, a debate that this chapter does not take up, but that is nonetheless indicative of one theoretical issue involved in assessing animality and human–animal relations in "religion," when the term itself remains open to question.

Another issue concerns a narrative that has taken hold in animal studies, one that posits the history of world religions as a movement from peaceful coexistence between humans and animals and a perceived fluidity-hybridity between human and animal forms, to violent teachings and practices, particularly of human dominion over other animals. For instance, in their General Introduction to *The Animal Ethics Reader*, editors Susan J. Armstrong and Richard G. Botzler suggest that for the

hunter-gatherer stage of prehistory and even for the agriculture and animal husbandry stage that followed it, a respectful attitude toward animals
prevailed—they cite for example the prohibition of the mistreatment of
animals in chapter 125 of the Egyptian *Book of the Dead*—an attitude that
was not to last, as over time, religious belief systems "became increasingly hierarchical" (1). As ethicist and animal advocate Bernie Rollin
puts it, an ancient husbandry "contract" prevailed for centuries between
humans and animals, and the essence of this contract was care. This age-
old contract was not free of human self-interest, Rollin writes: as much
a prudential as an ethical doctrine, it recognized that as humans benefit,
so do animals, that the relationship between them is symbiotic (7–10).

Within animal studies literature, the "histories" that have been
offered in support of the idea that religious belief systems, at one time
eschewing human-animal boundaries, became increasingly hierarchical
over time, are typically brief and often based on generalizations. This
may not be surprising, given the relatively recent development of animal
studies, and of specialized scholarship in world religions directed specifically to the animal studies field, and yet, at least for some, conclusions
have already been definitively drawn. Christianity has fared particularly
poorly here. As a case in point, Peter Singer's *Animal Liberation*—the text
out of which, it is often said, critical animal studies emerged—covering
everything from "pre-Christian thought" (Greek antiquity and Judaism)
to contemporary Christianity within five pages, concludes the following:
"the basic position of the ancient Hebrew writings toward nonhumans"
can be derived from Genesis 9: 1–3, and can be summed up as "human
dominion" (188); the Greek position that became persuasive for Christianity can be summarized in Aristotle's words, taken from his *Politics*, that
nature "has made all animals for the sake of man" (189); skipping quickly
to the fourth century CE, and the emperor Constantine's conversion
to Christianity, the tradition challenged the killing of humans during
gladiatorial combats, but not the killing or torturing of wild animals
(192); in keeping with this, Christianity in Roman times extinguished
"the spark of a wide compassion" that "a tiny number of more gentle
people"—Romans such as Ovid, Seneca, Porphyry, and Plutarch—had
kept alight (192–193); "We have to wait nearly sixteen hundred years"
after Constantine "before any Christian writer attacks cruelty to animals
with similar emphasis and detail on any ground other than that it may
encourage a tendency toward cruelty to humans" (193); skipping to the

Renaissance, Singer reaches Descartes, where Christian "dominion" (the word taken from Genesis) reaches its "absolute nadir" (200).

Similarly, in *Animals and Why They Matter*, Mary Midgley states quite simply that Christianity dismissed animals out of hand (12). Jim Mason in *An Unnatural Order: The Roots of Our Destruction of Nature* refers to "the West's aggressive and rigidly monotheistic Judeo-Christian-Islamic Megareligion" (30), reducing centuries of the histories of these three very distinct and diverse traditions into an abstraction not unlike "the animal." If we take it as a principle that differences matter and cannot (and should not) be subsumed uncritically into a supposed unity, then Mason's argument, in a chapter titled "Dominionism Identified," that "most of the world, unfortunately, has become infected with Western culture and its dominant ideas, even if all have not adopted the Judeo-Christian-Islamic religion" (30), might suggest the need for more research to be done regarding the legacy of world religions on "the animal" question. In the meantime, this chapter, sampling some of the available research on selected world religions, focuses on ethical teachings pertaining to the killing, sacrificing, and eating of animals.

Turning East

When philosopher G. W. F. Hegel penned his 1827 *Lectures on the Philosophy of Religion*, charting the east-to-west return journey of Spirit to the Absolute, he had little good to say about "primitive" "Eastern" (e.g., Egyptian and Indian) religions, Hinduism, for instance, where "there is no higher self-feeling or self-awareness present," and where "thinking has slipped back so wholly into abstraction" (289). Indicative of what some would call a *postmodern* sensibility, the 1970s and 1980s all but reversed the Hegelian course, declaring "Eastern" religions to be eminently more "spiritual" than those emanating from the decadent West, and advocating a turn back to their deep insights and ethical values. Physicist Fritjof Capra, author of *The Tao of Physics* (1975) and *The Turning Point* (1982), both international bestsellers, was one of many who found in the "new physics" (twentieth-century quantum mechanics and relativity theory) holistic and ecological views similar to those of Eastern mystics—and profoundly different from the still prevalent mechanistic conceptions of Descartes and Newton (see also Zukav; Pirsig; Heisenberg; Talbot). The wave of such writing that emerged from this period may well be an instance of what Edward Said described in his 1979 book *Orientalism* as the kind of research that imposes a romanticized, colonialist agenda on "Eastern" traditions—but it did nonetheless spark a Western fascination with "exotic" Asian religions.

At least to some extent, an idealized version of Eastern religions has made its way into animal studies, evident in statements as to the preference of Eastern over Western traditions, in references to the compassion for all sentient beings that lies at the core of a tradition such as Buddhism, and to vegetarianism as one of that tradition's moral requirements (see Singer, "Protection" 617).[1] In what follows, however, considering all-too-briefly traditions of Indian origin only, we refer to some studies indicating that generalizing statements about "Eastern" traditions (whether a single tradition or a number of traditions grouped under that label) obscure the complexity and diversity of their philosophical and ethical teachings, at least some of which are speciesist and anthropocentric, like their Western counterparts, and not all of which endorse vegetarianism.

Anthropocentrism and Speciesism

Hinduism, a complex tradition of many traditions and one of the oldest of world religions, originated in northwest India around 1500 BCE when a central Asian people who called themselves "Aryans" invaded the region, intermingling with an indigenous Harappan culture and developing the Vedic religion out of which the earliest Hindu scriptures were formed. The four Veda—canonical source texts for the first of many ethical traditions that historically and scripturally comprise what we now call "Hinduism," and widely regarded among Hindus as embodying the central truths of the tradition (de Bary 2)—have a strong ceremonial component centering on the performance of rituals of sacrifice, including animal sacrifice, under the authority of the priestly caste of brahmins. Vedic religion was hierarchically structured, led by male poets and priests, and androcentric—for example, the cosmic being in the Veda is male, Purusha, out of whose sacrifice creation is born (6). With the transition to Upanishadic Hinduism around 600 BCE, the tradition became more reflective and speculative, more interested in the symbolic meaning of ritual than in its performance (24), yet as Theodore de Bary points out, the texts called the Upanishads cannot be regarded as present-ing "a consistent, homogeneous, or unified philosophical system," and "[d]ivergences of method, opinion, and conclusion, are everywhere apparent within a single Upanishad" (24). It is fair to say, then, that even from the formative tradition alone, the Veda and Upanishads, no uni-fied "Hinduism" emerges, certainly none that is free of anthropocentric hierarchies or the slaughtering of animals. Rather, as Wendy Doniger suggests, in approaching the disparate parts of Hinduism, what we need is something like a Venn diagram, "a set of intersecting circles of concepts and beliefs, some of which were held by some Hindus, others by other Hindus, and still others shared not only by Hindus but also by believers in other South Asian religions, such as Buddhism or Jainism" ("Hindu-ism," 36). And as she goes on to note, "We would need a similar Venn diagram to do justice to Christianity or Judaism; religions are messy" (35–36).

Typically, Buddhism strikes Westerners as less "messy" than some others, particularly concerning its teachings about animality and human-animal relations. The tradition arose near Bihar in northern India in the sixth century BCE as an offshoot of Hinduism, and in part as a reac-tion against Hindu caste structure and blood sacrifice. In the scriptural texts, which were not compiled in written form for some five hundred years, Buddha refers to his teaching as the "Middle Way" to liberation, a moral path that avoids the extremes of either asceticism or worldliness, and therefore that does not denounce meat-eating (Buddha ate meat himself). At the same time, the first of Buddhism's Five Precepts is the injunction to "Refrain from taking life," a moral rule that is undergirded by the principle of *ahimsa*, or noninjury. Centuries of interpretation have been given to this moral precept and its underlying principle, but per-haps none is more authoritative for the tradition than that of the fifth-century monk and scriptural commentator, Buddhaghosa, who writes in his *Visuddhimagga (Path to Purification)* that the first precept disallows discrimination between "human" and "animal" life. "'Taking life' means to murder anything that lives," he writes. "It refers to the striking and killing of living beings," or of what might be referred to philosophically as "anything that has the life-force" (qtd. in Conze, *Buddhist*, 70–71). And as he goes on to explain, the first precept involves not only "negative morality" on the level of conduct—abstention from the action of kill-ing anything that lives—but also the kind of "positive" changes through which one abandons "the wish to kill" altogether, along with all violent or aggressive attitudes toward the other. The important role of meditation in Buddhism, and in Buddhaghosa's *Visuddhimagga*, has to do with this work of ridding oneself of self-centered same/different ways of thinking, and of internalizing instead such attitudes as "loving kindness." Indeed, one could say that, in Buddhism, morality (*sila*), which includes right action or right conduct, is not all it takes to stop the suffering (*dukkha*) that inevitably results from self-centeredness or craving for self (what the tradition calls "I-ing" and "mine-ing"). One needs also, through medita- tion or concentration (*samadhi*), to internalize right ethical principles and attitudes. Finally, to win liberation from ego-centeredness, one must gain wisdom (*prajna*) or insight into "the way things really are," into the "worldview" of *patica-samuppada*, the view that everything is connected "*Interbeing*" to everything else.[2]

Nevertheless, in a recent contribution to animal studies, *The Specter of Speciesism: Buddhist and Christian Views of Animals*, Paul Waldau of the Tufts School of Veterinary Medicine argues that formative Buddhism (the classical tradition as codified some five hundred years after Bud- dha's death in the Pali Canon) actually professes a negative and speciesist view of other animals—"*as other animals*" (154), differentiated as a class from humans and regarded as inferior to them. Waldau foregrounds what he calls "overtly exclusivist and anthropocentric features of the tradi- tion" (152), and, as part of his detailed study, even makes the case for the concept of dominion in Buddhism, suggesting that "the early Bud- dhists accepted human dominance of other animals when humans could, for their own benefit, effect such" (155). For Waldau, in short, classical Buddhism is speciesist, anthropocentric, and amenable to dominion over animals for reasons of human self-interest. This argument, which unfolds over two chapters, is based in large part on his reading of Buddhist *Theravada* canonical and extra-canonical texts: the *suttas*, or "sayings of the Bud- *13.* dha," the *Vinaya* texts that spell out moral rules for nuns and monks in the order, the more philosophical *Abhidamma* texts, the *Jatakas*, and other collections. Along the way, Waldau faults the tradition on a number of other points, suggesting, for example, that early Buddhists did not know some animals well and thus sometimes portrayed them "inaccurately" in their legends and stories, that is, with traits that "are not those of the natural world counterparts" (151). While arguing, for another example, that "Buddhists simply coexisted with daily, obvious harms to nonhu- mans" (149), he does not consider the differentiation within the tradi- tion of householders from monks and nuns, the latter regarded as more offer in service

advanced on the path to liberation, having removed themselves from
those worldly activities that are somehow bound up with doing injury or
taking life. The traditional insight here would be that one cannot live in
the world without doing injury to other living beings, and that, in order
to approximate the ideal of ahimsa, ascetic withdrawal and embracing of
the demanding ethics laid out in the *Vinaya* texts, are usually necessary.

Buddhaghosa explains in his *Visuddhimagga* that the rule against taking life applies more stringently to larger, than to smaller, animals, in part
because of the effort and intensity involved in the former case. Ian Harris
takes up this qualification of the First Precept in "'A vast unsupervised
recycling plant': Animals and the Buddhist Canon," suggesting that the
situation regarding noninjury is more complex than might at first appear.
"All of the ancient Indian renunciant traditions accepted the existence
of minuscule entities, but the Buddha's position was that 'if you can't
really see them, then you can't be said to have caused intentional harm.'
Buddhism, then steers a middle way between the inordinate diligence
of the Jains and a complete lack of care" (209). For Buddhism, as well,
killing an elephant is worse than killing a dog, for large animals require
more effort to kill, "and the degree of sustained intention must be con-
sequently greater" (209). Discussing the place animals hold in Buddhist
scriptures, Harris notes that although animals are considered to possess
the faculty of thought, "their ability to develop useful insights into the
true nature of things is limited" (208). Similarly, animals are less favorably
oriented to the possibility of liberation than are humans; they may be
constitutionally disposed to acts of violence and sexual misconduct; they
tend to disregard taboos held binding on humans; they are often classified alongside "human matricides, parricides, hermaphrodites, thieves,
and Buddha-killers" (208). Although Buddhism grants humans "a *gati* to
themselves, all animals are lumped together in a single category" (208).
Still, as Harris points out, Buddha spoke out against the immolation of
animals in Vedic sacrificial rites (209), and the tradition overall encourages kindness toward animals (213).

excessive, unusual

Even the Jain tradition, with what Harris calls its "inordinate diligence" concerning noninjury, betrays its own hierarchical ordering, albeit
one based on equality. This is one point Christopher Chapple makes
in "Inherent Value without Nostalgia: Animals and the Jaina Tradition,"
where he notes that, while the tradition teaches that each living being
houses a life force, or *jiva*, an apparent teaching of equality and of all life
as inviolable, the taxonomy of Jainism ranks birds, reptiles, and mammals

in the highest of an ascending five levels (241–242). The tradition considers human birth as the highest birth, and the human realm as the only one through which one might enter liberation (248). Yet, despite the somewhat ambivalent attitude toward animals taken in its stories, Jainism "seeks to uphold and respect animals as being fundamentally in reality not different from ourselves" (248), and it has a long tradition of developing animal shelters. Jains regard Mahavira (ca. 540–468 BCE) as their founder. A contemporary of Buddha, Mahavira—whose birth was presaged by his mother's animal dreams, and whose magnificent palanquin was decorated with animal images (243–244)—adopted an uncompromising adherence to *ahimsa*, which entails obedience to a number of rules regarding noninjury to plant and animal life, and regular fasting.

Studies of speciesism and anthropocentrism in the traditions of Judaism, Christianity, and Islam evidence a range of opinion, from wholesale damning of the three traditions as representing a unified, hierarchically ordered and violent "mega-religion" (Mason) to more apologetic and confessional efforts at recovering from them teachings of compassion toward animals. Careful scholars acknowledge that Western religions, like their Eastern counterparts, do not speak with a single voice. In "Hierarchy, Kinship, and Responsibility," for instance, Roberta Kalechofsky outlines a number of rabbinic and other interpretations concerning the Jewish relationship to the animal world, interpretations that vary on such important questions as whether humans have an absolute, or only a relative, duty to animals. Despite these varying interpretations, Kalechofsky suggests the following consensus teachings: Judaism posits a hierarchical scheme; with respect to animals, the tradition teaches kinship and compassion, but not reverence and not equality (92); Jewish values regarding animal life are based on quasi-equality or, approaching the modern era, even on inequality (94); Jewish teachings and laws regarding human responsibility for animals are embedded in family-modeled concepts of "hierarchy" and "dominion," but dominion is always limited (94); Judaism does not forbid the eating of meat, but does prescribe ritual manners by which animals can be killed (96); recently, some rabbis have accepted the shackling and hoisting of animals and even "the evils of factory farming" (96–97; see also Slifkin; Issacs).

Perhaps the first serious reassessments of Christianity's attitudes toward nonhuman life were prompted by growing awareness, during the late 1960s and 1970s, of a looming environmental crisis. Paul Santmire's *The Travail of Nature: The Ambiguous Ecological Promise of Christian Theology* comes out of this period, and although its concern is with Christian attitudes toward nature, rather than toward nonhuman animals specifically, it does analyze the anthropocentric ideas and claims to human dominion that became prominent during the Reformation and early modern periods, in John Calvin, for example, with whom anthropocentrism and dominion take on a new dynamism. With Calvin, Santmire writes, "[h]uman dominion over history and the created order generally now receives a powerful theological validation" (126). As Keith Thomas puts it in *Man and the Natural World: Changing Attitudes in England 1500–1800*, Christianity provided "the moral underpinnings for that ascendancy of man over nature which had by the early modern period become the accepted goal of human endeavour" (22).

Although Paul Waldau would agree, he views the notion of dominion as belonging to the Christian tradition overall, rather than specifically to its early-modern or modern periods. Indeed, in *The Specter of Speciesism*, he suggests that all three terms—*anthropocentrism, dominion,* and *speciesism*—apply to Christianity's understanding of animals. He traces anthropocentrism and dominion right back to the Hebrew Bible, and while allowing for differing interpretations of traditional views (and taking particular issue with the interpretations of Andrew Linzey), he finds what he sometimes refers to reductively as the "Judeo-Christian" tradition to espouse both anthropocentrism and human dominion over the nonhuman world (202–211). In addition, he argues that "the mainline tradition" has values, emphases and exclusions that fit his own definition of speciesism, that is, "the inclusion of all human animals within, and the exclusion of all other animals from, the moral circle" (38; 216). It might be important to read Waldau's study of the overall failures of Buddhism and Christianity to offer an "inclusive" animal ethics alongside his own argument that traditional ethical teachings with respect to animals should be evaluated on the basis of a "like-us" standard. He focuses on the ethical consideration that traditions give to what he calls "key species," defining those as "by far the most closely related to humans" (61), thus marked by "large brains, communications between individuals, prolonged periods of development in complex familial and social envelopes, and levels of both social integration and individuality that humans can recognize" (60).

Although focused studies are emerging of animality and animal welfare in Islam (Masri; Abbas; Foltz), much research remains to be done. Generalizing statements, favorable and unfavorable, are not uncommon. As an example of the latter, Katherine Wills Perlo in *Kinship and Killing,* writes that "Dominion over animals is seen as one of the benefits to human beings in a scheme of things to which they must submit for good or ill; and their own obedience to Allah—who, like the stockbreeder of the Jewish Bible gives life and death (45:26, 53:44) and determines ageing and the time of death (40:67)— justifies the submission which they forcibly extract from animals" (98). More carefully, Martin Forward and Mohamed Alam point out that although human beings have power over animals in Islam and can use animals for human purposes, this power comes with qualifications: animals are not to be caged, beaten unnecessarily, branded on the face, or allowed to fight each other for human entertainment; vivisection is forbidden; the tradition opposes battery farming and cruel methods of animal husbandry. In short, Muslims have moral obligations toward animals (294).

Eating Animals

We might begin this section, as we have elsewhere in this book, with reference to Peter Singer and Tom Regan, whose work provides something of a benchmark for many issues in critical animal studies. Regan's position on eating animals is clear, and has not wavered since he published *The Case for Animal Rights*: as he puts it in that book, "Vegetarianism is not supererogatory; it is obligatory" (346). Raising animals for food, whether by traditional or factory farm methods, is unjust (345). As well, to purchase animal products is to participate in an unjust practice (consumers would be justified in buying meat only if the farming practices involved "treated animals with the respect they are due" (346). For the most part, Singer agrees with Regan, arguing in *Animal Liberation* that becoming a vegetarian is one of the most practical and effective steps one can take toward ending the killing of nonhuman animals and the infliction of suffering upon them (161). At the very least, consumers should be sure of the origin of the particular items they purchase, avoiding animal products produced by factory farms (170). Although he has continued over the years to advocate vegetarian and vegan lifestyles, Singer has also conceded that most of the hundreds of millions of people who live in industrialized countries are unlikely to go vegan any time soon, which is why, as we have noted, in a book like *The Ethics of What We Eat*, co-authored with Jim Mason, he outlines practical methods that even meat-eaters can adopt for reduction of animal suffering (see 278–280).

A number of ethicists follow Singer in qualifying the moral requirement of vegetarianism. For instance, James Rachels in "The Basic Argument for Vegetarianism," focuses on the immorality of today's meat-production business, which involves too much animal suffering without good enough reason to outweigh it (we can get along without eating factory farm products); and David DeGrazia in "Animals for Food," although not necessarily ruling out meat-eating or killing, argues that purchasing factory-farmed meat is morally indefensible, given the massive suffering industrialized farming inflicts on animals.

With the exception of Jainism, the world religions we approach in this chapter are closer to Singer than to Regan on the issue of vegetarianism—or at least, such is the conclusion suggested by recent interpretations. The traditions of Judaism, Christianity, Islam, Hinduism, and Buddhism do not condemn meat-eating outright—although some animal ethicists argue that vegetarianism may be regarded as realizing more fully the nonviolent ethical ideal espoused by these traditions. This is the interpretation of Christianity that Andrew Linzey offers in "The Bible and Killing for Food," an essay that demonstrates some of the difficulties involved in translating scriptural writings into contemporary ethical teachings. Linzey's study concerns two passages from the Book of Genesis, the first (Genesis 1) suggesting vegetarianism (God giving "green plants for food" to "everything that has the breath of life") and the second (Genesis 9) seeming to reverse this (God giving to Noah and his sons for food "every moving thing that lives"). Linzey reconciles this apparent contradiction by noting, first, that although the Genesis texts originated out of a non-vegetarian culture, we can deduce from the descriptions of paradise in these texts that early Hebrews "were deeply convinced of the view that violence between humans and animals, and indeed between animal species themselves, was not God's original will for creation" (287). Similarly, he reads the Noah story as a portrayal of human corruption and sinfulness, rather than as a license to kill and eat meat. If God allows killing for food after the Flood, this is only in situations of necessity, Linzey maintains, and only with the proviso that, "for every life you kill you are personally accountable to God" (288). Moving to Isaiah 11, he suggests further that, while biblical writers sometimes considered killing to be justifiable in the present time, they also insisted that, with the arrival of another Messianic Age, peaceful, paradisal relations would again be restored (288).

By appealing to both creation story origins and messianic end times, Linzey draws from the Hebrew Bible perspectives on killing for food that, he says, not only display "internal integrity" but that also have "enormous relevance" to current debates concerning animal rights and vegetarianism (289). He argues that the Bible does not minimize the gravity of killing animals any more than it condones killing as God's will. Moreover, in his view, those who wish to adopt a vegetarian or vegan lifestyle "have solid biblical support" (289).[3] The question is not, he writes, whether killing animals has *never been* justifiable, but rather whether it is necessary *now*, when it is not essential to kill in order to live, and when it is "perfectly possible to sustain a healthy diet without any recourse to flesh products" (289). In short, Linzey regards the vegetarian lifestyle closer than the carnivorous one to the biblical ideal of peacefulness (289). In additional support for this view, he concludes that while Jesus ate fish—in the context of first-century Palestine, where geographical factors alone suggest a scarcity of protein, there may have been a real need so to do—his overall message is one of peace, a message informed by the messianic ideal of reconciliation and harmony (291–292).

Research on vegetarianism in world religions is vexed by apologetic publications such as Steven Rosen's *Diet for Transcendence*, written apparently for "believers," and accrediting a particular "truth" to traditional religious teachings, as if they directly represent "the will of God." Rosen maintains that in each of the traditions he considers—Judaism, Christianity, Islam, Buddhism, and Hinduism—there is much to endorse vegetarianism, arguing moreover that, "With the mounting scientific evidence that a meatless diet offers a more healthful life, and that eating flesh shortens one's lifespan, reason dictates that God would choose a vegetarian diet for His children" (15). *Religious Vegetarianism: From Hesiod to the Dalai Lama*, edited by Kerry Walters and Lisa Portmess, combines devotional and interpretive material. Rynn Berry's *Food for the Gods: Vegetarianism and the World's Religions* combines recipes, interviews, and essays on vegetarianism and world religions.

Christopher Key Chapple points out in "Inherent Value Without Nostalgia" that all Jain adherents are expected to abstain from eating animal flesh; that laypeople are expected to avoid professions that harm animals directly or indirectly; and that monks and nuns strive to minimize violence even to microorganisms and plants (247–248). In *Nonviolence to Animals, Earth, and Self in Asian Traditions*, Chapple further describes Jain

fasting practices, including the fast unto death, the latter considered to be
the most auspicious way that life can end. The recognition behind this
practice is that violence is necessary to survival, that eating necessarily
involves the taking of life, even if vegetable life (99). Yet the Jain final fast
is undertaken only if specific criteria are met, and it is distinguished from
suicide, which the tradition condemns (101–102).

Animal Sacrifice

That animal sacrifice has precedence in both Eastern and Western religions is not in dispute. Interpretation of traditional sacrificial practices
and rituals, and particularly of their relevance for contemporary animal
ethics, betrays less of a consensus. Thus, for example, although Andrew
Linzey suggests in *Christianity and the Rights of Animals* that the ancient
Hebrew tradition of animal sacrifice did not necessarily involve a low
view of animal life (41), Paul Waldau in *The Specter of Speciesism* takes
issue with this view, suggesting that "any argument that humans' sacrificing other animals is a confirmation of those animals' importance is para-
sitic on an underlying notion that humans are more important, and that
their God affirms this by accepting the sacrifice of another living thing
by humans" (211). In animal sacrifice, Waldau insists, "the animal dies for
human interests, not for its own" (211–212).

Kimberley Patton advances just the opposite view. In "Animal Sacrifice: Metaphysics of the Sublimated Victim," taking issue with Tom
Regan's *Animal Sacrifices*, which uses the term *sacrifice* with reference to
animal experimentations in which the animal does not survive the procedure, she argues that traditionally in religious animal sacrifice, animals,
"far from being *things*" are seen as "active subjects from start to finish in
the sacrificial ritual" (392–393). Ritual sacrifice, given its "metaphysical
situation," cannot be compared with the use of animals in science labs or
clinical trials; it is "not the same as ordinary killing, and has never been"
(393); and it actually hallows and empowers animals (402). To make this
case, Patton interprets traditions of animal sacrifice under four head-
ings: required perfection and ritual beautification of the victim; voluntary cooperation of the animal (rather than simply killing the restrained
victim, ritual sacrifice "requires that the animal first 'assent' to its own
destruction"); religious elevation and individuation of the animal victim (sacrifice "rescues" the victim "from inconsequentiality as one of a

multitudinous herd"); eschatological dimensions of the sacrificed animal ("its temporary divinized state, effected from the moment of selection for sacrifice and emphasized by the ritual killing itself, is often rendered permanent, and its eschatological future assured in a kind of glistering light") (394–401).[4]

Whether the ritual sacrificial logic Patton describes is "false" is open, she says, to question, but the logic itself must be understood, she maintains, before this question can be addressed (402–403). Peter Singer seems to take issue with such framing of animal sacrifice in *Animal Liberation*, where he makes the point that the "logic" animal sacrifice once followed may well be overturned by contemporary adherence to ritual rules. For example, he notes that orthodox Jewish and Islamic dietary laws forbid the consumption of meat from an animal who is not "healthy and moving" when killed, a rule that may have been intended to prohibit the eating of an animal found sick or dead, and that—by specifying that the killing be carried out with a single cut with a sharp knife aimed at the jugular and carotid veins—may have been a more humane method of killing than any available alternative (153). Today, however, this rule "is less humane, under the best circumstances, than, for example, the use of a captive-bolt pistol to render an animal instantly insensible" (153–154). After reading Singer's outline of the shackling and hoisting of conscious animals beings ritually slaughtered, most would agree.

Scholars of religion will continue to contribute analyses of ritual slaughter to the animal studies field. And while Kimberley Patton makes light of the social and gender structures that, some argue, are inscribed through rituals of animal sacrifice performed by male religious elites, these structures, too, are being disclosed and debated. Carol Adams, for example, in "The Rape of Animals, The Butchering of Women," points to the common oppression of women and animals in meat-eating Western culture, and to the fragmentation and renaming of body parts subsequent to slaughter, when "cows become roast beef, steak, hamburger; pigs become pork, bacon, sausage" (272). Only after covering animals with sauce and disguising their original nature, she argues, can consumption occur. We will return to the matter of "consuming the other" in the closing chapter of this book.

9

THE SUBJECT OF ETHICS

Cultural Studies, Art, Architecture, and Literature

Opening her 2009 book, *Stalking the Subject*, professor of English Carrie Rohman writes that: "Literary studies and critical theory are witnessing the development of a new discipline surrounding the cultural and discursive significance of animality and its relationship to metaphysics and humanist discourses. Whether this discipline becomes known as 'critical animal studies,' or the simpler 'animal studies,' the contours of the field are taking shape across broad fields of inquiry" (1). Here, Rohman clearly includes literature (literary studies and literary criticism) within critical animals studies, conceiving of the latter as reaching across disciplines. "From the strictly philosophical to the historical, cultural, and literary," she writes, "the past five years have brought about an unprecedented amount of scholarly work on the place, meaning, and ethical status of animals in relation to our signifying practices" (1). Concurring with Rohman in considering critical animal studies to be a broadly inclusive and multidisciplinary field, this chapter turns to literature, cultural studies, art, and architecture as contributing centrally to "the animal" question. Aside from the limited space available to me in an introductory book like this, one reason for grouping architecture, art, cultural studies, and literature together today is that their theoretical and scholarly endeavors frequently overlap, most notably, perhaps, as concerns "the subject of ethics"—that is, as concerns their questioning of the status both of the *subject* and of *ethics*.

Addressing critical animal studies from within the academic institution, Rohman refers to the field as "a new discipline." As we have noted, however, precisely for reason of its inclusivity and multidisciplinary nature, critical animal studies tends no longer to be delimited to an intact unit or "discipline" within today's research university; at the same time, literature, (including comparative literature), art, architecture, and cultural studies have been decisive in its development, with many historical and theoretical courses now embedded in their respective undergraduate and graduate curricula. The openness of these units to animal studies parallels their receptivity, since the 1960s, to continental philosophy and "theory," so-called poststructuralist critical theory included, all of which could find no home in philosophy departments dominated by Anglo-American approaches. Oddly enough, by the end of the twentieth century, in most North American universities, continental philosophy was being taught outside of philosophy departments; cultural studies had emerged as a significant critical and creative, at once philosophical and literary, pursuit; architecture had been marked for decades by its engagement with "theory"; literary criticism had come to include as much philosophy as literature; and visual arts studies had long been exploring "postmodern" and "poststructuralist" theoretical-philosophical approaches.

In this chapter, my purpose is not to discuss this remarkable, boundary-crossing development within the academic institution, nor to survey the vast realms of literature, cultural studies, art, and architecture, each rich in its own right where the question of "the animal" is concerned. Rather, my more limited goal is to focus on two questions. The first question, sparked by the strong opinion we have noted within the field that critical animals studies is properly philosophical, in the Anglo-American sense, and thus not the purview of the literary and visual arts (and this would surely include architecture and cultural studies, both of which entail both critical and creative pursuits), is: Why should literature, cultural studies, art, and architecture have a place within critical animal studies? The second question is: How do these "extra-philosophical" pursuits contribute to a *critical* animal studies, particularly to much-needed *philosophical* questioning of the contemporary status both of the *subject* and of *ethics*? In what follows, guided by these questions, my discussion, much confined for reasons of space, proceeds with reference only to selected example-texts.

Cultural Studies

The relatively recent realm of inquiry called *cultural studies* does not "belong" to literature, although it is often broached from within academic literature units—or, depending on the specific academic institution in question, the study of "literature" (English literature), comparative literature (world literature), literary or critical theory, and cultural studies, might be taught within the same unit or critical-creative cluster of units, with animal studies courses found in any or all of these. A number of today's religion departments do cultural studies work as well, and could be labeled "religion and cultural studies" units. My purpose here, however, is not to imply any particular institutional placement for cultural studies, but rather to suggest, first, that cultural studies is one scholarly area that crosses traditional academic boundaries; and second, that it is, and will continue to be, an indispensable site for critical animal studies. As Canadian cultural studies scholar Jodey Castricano notes in her edited collection, *Animal Subjects: An Ethical Reader in a Posthuman World*, one of the difficulties of locating cultural studies within any humanist discipline is that the field, still in transition, continues to move beyond the boundaries of traditional humanism, notably for reason of its own self-critique, "the questions it asks of its own relation to power" (4; see the discussion below of Mieke Bal's argument on the necessity of such questioning). In addition, Castricano suggests, cultural studies exceeds the confines of humanism in breaching anthropocentric borders, questioning "the limitations of traditional humanist philosophy that is concerned primarily with the welfare of humankind to the extent that the faith of the humanist in empathy and democracy is steeped in a host of anthropocentrisms" (5). Defining cultural studies this way—not as discipline-based but as characterized by posthumanist engagement—explains the inclusion in Castricano's book of contributions from lawyer Lesli Bisgould, biologist Ann Innis Dagg, and literary critic Cary Wolfe, among others. In calling into question the limitations of traditional humanist philosophy, a collection such as *Animal Subjects* would involve, necessarily—if in this case not uniformly—a critique of the humanist subject or self, and of an animal ethics that is grounded in a subject-centered standard.

That cultural studies also embraces art and architecture is evident in the work of Mieke Bal. During her career as theory of literature professor at the University of Amsterdam and founder of that university's School for Cultural Analysis, Bal published numerous books and essays in inter-disciplinary cultural analysis, visual art, narrative, and cultural memory, all of which in some way contend with subjectivity and/in representation, or with what, in the subtitle of her book, *Double Exposures*, she refers to as: *The Subject of Cultural Analysis*. In this text, Bal turns to the museum—real museums, and the metaphorical use of the idea of "museum"—as what she calls "a suitable emblem of contemporary humanistic studies" (2). In other words, she takes the museum to be "the place" where we might find the assumptions of humanism "on display," including assumptions as to the status of the human subject and its relation to nonhuman animals. *Display* is a key word here. Museums are houses of display, of what Bal terms *exposition* or "exposé." They are involved in exposing and showing something to a public, in gestures of exposé that, for Bal, are never neutral, always culturally and ethically coded. In *Double Exposures*, she undertakes a critical analysis not of the museum as an object, but of the "discourse" or "gesture" of showing that is deployed in and through it: in her approach, museum exposure is "double," then, exposing human-ist assumptions in the way it puts "things" on display (3).

To follow Bal's analysis, readers have to let go of certain notions, for example, that "discourse" is confined to written texts, or that art and architecture—museums, for example—are empty of ethical and politi-cal arguments. Indeed, she approaches museum displays as "discursive" acts, performances analogous to speech acts, through which authoritative (humanist) beliefs are affirmed. For her, every act of exposing or showing involves an agent or subject, who "puts 'things' on display, which creates a subject/object dichotomy. This dichotomy enables the subject to make a statement about the object" (3). Staying with the speech act analogy, Bal refers to the subject doing the showing as the first-person "addres-sor" and the viewer (the people who visit the museum) as the second person "addressee." In a museum exhibition, "a 'first person,' the exposer, tells a 'second person,' the visitor, about a 'third person,' the object on display, which does not participate in the conversation" (3–4). This dis-course of display has a truth value, Bal notes: it is as if the first person (who remains invisible) were saying to the second person (visitor) about the third person (the mute object on exhibit), "Look! That's how it is" (4–5). This is the museum model of expository discourse, but it is also the humanist model for all "rational" discourse, including the discourse of "rational ethics," where the speaking subject remains "invisible," that is, "hides" behind the realist claim to universal fact or truth. In cultural studies texts such as Bal's, this discourse of exposé is exposed, and its assumptions brought into question. That said, Bal also recognizes that there is a dual exposure involved in every expository discourse, including hers, "an inevitable duplicity within, and involvement of, the field out of which this study grows" (7).

To briefly clarify Bal's analysis of expository discourse, we might follow her into New York's American Museum of Natural History. Here, she suggests, the "I," the expository agent who is "speaking" in the museum, belongs to "an era of scientific and colonial ambition, stretching out from the Renaissance through the early twentieth century" (17). The American Museum of Natural History is thus saddled with a double status, being at once a museum, and a museum of a museum, "a reservation, not for endangered natural species but for an endangered cultural self, a meta-museum" (17). We can appreciate this double status in the darkened Hall of Asian Mammals, for example, where the visitor is surrounded by dioramas of exotic animals one might know only from postcards and geography books, strange animals placed against painted backdrops, the dioramas illuminated from within, an effect that "highlights the object while obscuring the subject" who has constructed this display (20). Bal notes that a door at the end of this hall opens to the Hall of Asian Peoples, effecting a problematic transition from mammals to peoples: "The most obvious problem is the juxtaposition of animals and human foreign cultures" (22), as if, for the visitor, both belong to an otherness that, in this realist setting, is depicted as "natural."

The transition between the two halls "is monitored by a small display, within the Hall of Asian Mammals, at the far end to the right, between the Indian rhinoceros on the left and the water buffalo on the right": a nineteenth-century statue from Nepal of "Queen Maya giving birth to the Buddha from her side," as the result of a visit by a white elephant with golden tusks (22). Bal notes that the panel underneath the statue anchors it in "popular tradition," emphasizes Buddhism's polytheistic tendency (in contrast with Christian monotheism), further estranges the tradition from the Western viewer by using the word "sects," and attributes Buddhist "vegetarianism" to another strange feature of the tradition, with its "many stories of the previous lives of the Buddha as a compassionate soul in the bodies of animals" (24). What particularly Western vision informs placement of the statue of a human, Queen Maya, in a hall where animals are on display? In this case, Bal suggests, both gender politics and ethnic stereotyping are behind the "Look! That's how it is" exposé. For the panel beneath the statue explains the relevance, for Buddhists, of human presence in the animal realm, but does so "by qualifying the humans in question as being close to animals, closer than 'we' are," and it adds an element of the foreign and exotic by portraying human conception from an animal (25). "The statue represents a woman, 'naturally' as close to animals as humans come, and it represents her in the most 'natural' of poses, in giving birth," a selected representation of femininity that "affirms woman's closeness to nature through an unnatural fiction presented as foreign other" (25). And with this "surplus of ideological information that the panel with the verbal representation is conjured up to 'explain'" (26), the visitor, Bal says, is now prepared to enter the next hall.

Introducing the visitor to the Hall of Asian Peoples is a panel displaying the headline, "Man's Rise to Civilization." After discussing some implications of the words, *man*, *rise*, and *civilization*, Bal walks her reader through the hall's portrayal of an ascendancy from "primitive" cultures to "our own" high civilization, anonymity yielding along the way to "great names," such that the transition being charted in this particular hall is not between animals and foreign peoples, but "between the latter and 'us'" (28–29). In short, Bal suggests, "[t]he peoples represented in the American Museum of Natural History, just like the animals, belong irreducibly to 'them': the other, constructed in representation by an 'I,' the expository agent. The 'you' for whose benefit the 'third person,' here the Asian peoples, is represented, is constructed by implication [. . .] as belonging to the Western white hegemonic culture" (30). There is much more to the museum's "rhetorical strategy," as Bal analyzes it in these early pages of *Double Exposures*: her reading of a panel in the Hall of Asian Peoples portraying nineteenth-century Siberian dog sacrifice, for example, and subsequent to that, her journey through the Hall of African peoples. Her interest throughout is to probe ways in which, in each display, verbal and visual messages work together to present a realistic narrative that hierarchizes "us" (civilized, white, Christian, males) in relation to "them" (foreigners, females, animals).

But the book is more than an exposé of the hidden assumptions behind the "speaking I," in this case, the museum as expository agent: it is also replete with suggestions, both visual and verbal, as to how a first-person subject, an "I"—whether the addressor-agent be a museum curator or an ethicist writing on animal issues—might make him- or herself more visible, more readable in his or her representations, and in the process might open the taxonomy of "us" and "them" (human/animal, man/woman, civilization/savage, culture/nature, same/different) to self-critique. To be effective, such critique must be sustained throughout exhibits, as throughout all other texts; it must be pursued "inside the hall," and not positioned as a preface outside the representational frame (37).

Why cultural studies? *Double Exposures*, with its probing of "the subject of cultural analysis," is already more than fifteen years old, yet the book remains ahead of much work in critical animal studies in recognizing that structures of oppositional difference are rooted in a metaphysics of the self; in making the case that this self is inextricably meshed with ethnic, gender, and species stereotyping; in noting that the "texts" through which this self speaks its "Look! That's how it is" include buildings as well as books, inscribed walls as well as iPads and pages; and that any discourse of exposition not open to self-critique, ethics included, can only perpetuate the subject/object, us/them, human/animal dichotomy.

Bal's writing overall demonstrates the importance of art and architecture to the field of critical animal studies, and certainly, some of her works could be included below as a "key" art and/or architecture texts. Her book-length treatment of Louise Bourgeois' 1997 work *Spider* is a case in point. For Bal, Bourgeois' drawings, installations, and enormous spider sculptures together comprise a kind of text, one that invokes a home, the skeleton of a house, that builds the sense of a habitat, although one in which the subject is not, or no longer, master. Through the *Spider* series—the spider, an insect few would recognize as having a claim to interests or rights—Bourgeois "turns the metaphor of the mind's house, whose master does not master it, into a literal, embodied work of architecture" (39). It is as if Bal reads in certain traditions of art and architecture a counter-culture, or counter-memory: in Rembrandt's painting of the *Slaughtered Ox*, for instance, a painting he executed at the same time that Descartes (Rembrandt's neighbor in Amsterdam) was writing his account of the *bête-machine*, but that, Bal suggests, loosens the boundaries of the body even as it disturbs the neat division between human and animal (*Reading*, 387).

Art and Architecture

Early in her book, *Architecture, Animal, Human: The Asymmetrical Condition*, professor of architecture Catherine Ingraham recalls her reading of Le Corbusier's *The City of Tomorrow*, with its claims that the existence of right angles and straight lines are primary evidence for the "rightness" of the human mind and the "uprightness" of architectural thinking. In making this claim, Le Corbusier opposed human life to animal life; the latter, he said, repeating a classical gesture, is not characterized by human uprightness. As Ingraham notes, he used the donkey as an example: "The wandering, mindless donkey, for Le Corbusier, was the instructive foil to the straightness of the lines that humans draw in the world. Thus, ancient cites organized around animal paths are, according to Le Corbusier, sites of congestion and disease, while modern cities exhibit their health by means of straight avenues based on right angles" (13). Ingraham goes on a few pages later to cite a comment made about Le Corbusier's example by one of her former students: "if you are riding a donkey or carrying things by donkey then the much-maligned 'donkey paths' that Le Corbusier proposes replacing with more 'rational' avenues are, in fact, the more rational of the two pathways" (16; see also Ingraham, *Linearity*). Nevertheless, if at some point in history, donkeys were allowed to move about as they chose, and if "donkey urbanism" was a possibility, this is not the case in what Ingraham calls our "post-animal" world, a world in which humans have left their animalness, and animals, behind (16).

Why architecture? As Ingraham's study suggests, the genealogy of this abandonment calls for sustained analysis from within critical animal studies, and from within architecture as contributing to the field. *Architecture, Animal, Human* examines three historical periods during which "certain terms/clusters of, in particular Western, ideas around architecture and life" are brought into "intimate relation to each other" (11). These periods include the Renaissance, during which "humanism" emerges; the Enlightenment, when "human life is definitively separated from other forms of biological life," and when, with the emergence of biology as a distinct discipline, architecture adopts the life sciences as its model; and finally, the contemporary period, when "theoretical work in architecture and biology, particularly genetics, coupled with computational technologies, is suggestively critiquing ancient divisions between the animate and the inanimate" (11). Needless to say, this critiquing of animate/inanimate divisions comes at a critical moment, and situates architecture, among other pursuits, at the forefront of current animal ethics and environmental sustainability discussions.

As I mention earlier in this chapter, for the past several decades contemporary visual art, like architecture, has been exploring poststructuralist and posthumanist critiques of the modern human subject, particularly as these might contribute to a reshaping of the traditional, anthropocentric man/animal dichotomy, and by implication, then, to a rethinking of animal ethics. According to Steve Baker in *The Postmodern Animal*, despite its complexity, postmodern art can be summarized as dealing with "the animal" question either by aligning itself with the work of conservationists and endorsing "animal life itself," or by using irony to call received wisdom into question (9). In either case, postmodern art betrays some skepticism about modernity's "operation of truth and knowledge" (12); tends to distance itself from the culture and institution of the "great artist," abandoning the studio in favor of "the field," in the process, disrupting the notion of the "originating" author (14); and invites "the participating of animals themselves in the mark-making" of art (14). Baker's text is philosophically informed, so much so that it defies any easy bifurcation of "art" (the creative) from "philosophy" (the critical). Drawing its examples from recent art *and* philosophy, *The Postmodern Animal* explores the role of irony in contemporary "animal" art, as well as in the philosophical, "becoming-animal" work of Gilles Deleuze and Felix Guattari (19; see also Baker, *Picturing*).

Taking some exception to studies such as Baker's that bring animals into art by way of irony, as if in fear of "familiarizing" or "sentimentalizing" them, Alice Kuzniar, in *Melancholia's Dog*, takes the "closeness" of animals—in this case pets, specifically dogs—to have everything to do with their ungraspable "otherness." Kuzniar's book is a remarkable study, as much philosophical as literary, given as much to visual art as to psychoanalytic theories of melancholia, a book that is impossible to "slot" within a single discipline, and that has much to contribute to critical animal studies. In Deleuze and Guattari's notion of "becoming animal," as in Baker's resort to irony, Kuzniar detects a "mistrust" of pets, "Oedipal animals each with its own petty history," as she cites Deleuze and Guattari who, akin to rights-theorist Tom Regan, warn, she says, "against the sensibilities of empathy for animals because they are insufficient for enacting change" (4). Unwilling to "distance" herself from "pets," or to deem consideration of them as peripheral to "serious" scholarly discourse, Kuzniar attempts to find a way of speaking about their very proximity as otherness. In doing so, she challenges traditional tenets concerning both the human subject, and the discourse of ethics. "It is not the dog who cannot speak and must learn our language (sit, down, stay), but we who cannot speak properly about it, whereby this speaking would not include the pretense of making it think silly human thoughts," she writes. The question is not how to project into human words what a dog thinks, "but how to preserve, respect, and meditate on its muteness and otherness" (2).

Why art? At once a critical work on visual art and an exercise in "canine cultural studies" (ix), *Melancholia's Dog*, in the spirit of Jacques Derrida, calls into question the human/animal binary, particularly as it is thwarted by the "closeness" of the animal, the pet (e.g., Derrida's cat, his little pet, whose challenge to ethics he discusses in the opening pages of *The Animal That Therefore I Am*), the pet as unique and "totally other" (5). In four chapters, the book considers muteness, shame, intimacy, and mourning, making crucial references to selected works of visual art: in one instance, for example, juxtaposing the (1985–1986 and 1986–1987) portrait paintings of (Sigmund Freud's grandson) Lucian Freud—paintings that portray a woman with her whippet(s), the woman and the hound(s) in close proximity to, and somehow doubles of, each other (12–14)—with the copper etching of *Melencholia I*, executed by Albrecht Dürer in 1514, again showing a woman in close proximity to her dog. "Double, melancholic portraits, then, lie here next to each other—Lucian Freud next to Albrecht Dürer, woman next to hound" (17). Here, as elsewhere in Kuzniar's book, what arises from such (philosophical-artistic) doubling is the question of how to read one painting through the other, or more precisely, how to read one woman—Freud's blue woman or Dürer's *Melencholia*—through her dog. For one thing, Kuzniar suggests, in both works of art, the stillness of the woman and of her brute attendant can be seen as a sign of silence, of a "stillness or speechlessness" that points back to "the limits of interpretation," the limits, we might say, of any attempt to "gauge boundaries correctly" (18).

Literature and Literary Criticism

J. M. Coetzee published his novel, *The Lives of Animals*, in 1999. An acclaimed novelist and literary critic, Coetzee received the Nobel Prize for Literature in 2003, and has twice been awarded the Booker Prize (1983, 1999). In 1997–1998, he was invited to give the Tanner Lectures at Princeton University, but rather than delivering two conventional lectures for the occasion, Coetzee presented parts I and II of *The Lives of Animals*, which is the fictional account of an aging female academic, Elizabeth Costello, who, it so happens, was herself invited to deliver lectures at an American university (the fictional Appleton College) where her son John is an assistant professor of physics and astronomy. As it turns out, the fictional lecturer, Elizabeth Costello, chooses to speak on the topic of human maltreatment of, and cruelty to, animals. Coetzee's novel centers on her two fictional lectures ("The Philosophers and the Animals" and "The Poets and the Animals"), which are set in the narrative of her relationship with John and his wife, and of her exchanges with academics and students at Appleton. *The Lives of Animals*, then, Coetzee's Tanner Lectures, comprises "two lectures within two lectures," as the political philosopher Amy Gutmann puts it in her Introduction to the book (1). In addition, four critical responses to Coetzee's Tanner Lectures are included in *The Lives of Animals*: these by Hindu scholar and religionist Wendy Doniger, whom we cite in the previous chapter; evolutionary theorist and primatologist Barbara Smuts; literary theorist Marjorie Garber; and moral philosopher Peter Singer, whose *Animal Liberation* and other works we refer to frequently in earlier chapters.

In brief summary, Coetzee's character, Elizabeth Costello (whom we meet again in his 1999 novel, *Elizabeth Costello*), an Australian writer known particularly for her 1969 novel, *The House on Eccles Street*, now gives her life largely to attending conferences around the world, delivering invited lectures and accepting various awards. Soon after she arrives at Waltham for the three-day visit during which she will deliver her two Gates Lectures, we learn that Elizabeth is not beloved by her daughter-in-law, Norma, and that the issue of meat-eating is one matter of contention between them. How passionate an issue vegetarianism is for Elizabeth becomes clear the next afternoon during her first lecture, which opens with reference to Franz Kafka's story, "Report to an Academy," and its character, Red Peter, an educated ape "who stands before the members of a learned society telling the story of his life—of his ascent from beast to something approaching man" (18). Confessing that she feels a little like Red Peter herself, Costello moves quickly to a contentious comparison between the horrors humans inflict on animals in abattoirs, trawlers, and laboratories all over the world, and Nazi crimes of genocide during World War II, the horrors of torture and death in Third Reich concentration camps such as Treblinka. In both cases, she says, humans aware of such horrors have shown themselves ready to turn a blind eye. Not Elizabeth Costello, who puts her point bluntly: "Let me say it openly: we are surrounded by an enterprise of degradation, cruelty, killing which rivals anything that the Third Reich was capable of, indeed dwarfs it, in that ours is an enterprise without end, self-regenerating, bringing rabbits, rats, poultry, livestock ceaselessly into the world for the purpose of killing them" (21).

Costello's intention, she says, is not to polarize people and incite a polemical exchange: she wants to facilitate a cool-headed "philosophical" discussion. Yet Western philosophy, from Aristotle through Descartes to the present, does not offer her much, that is not unless she wants to entertain debates as to whether animals have or do not have souls, whether they can reason or are simply biological automatons, and so on. Reason, she suggests, may be no more than "a certain spectrum of human thinking. And if this is so, if that is what I believe, then why should I bow to reason this afternoon and content myself with embroidering on the discourse of the old philosophers?" (23). Rather than entering debates as to whether animals have the capacity to reason or to suffer—debates she considers to be outworn and largely irrelevant—Costello returns in this first lecture to the fictional Red Peter and to another ape, Sultan, perhaps Red Peter's prototype, captured on the African mainland in the early 1900s and subjected to experiments that are described by the psychologist Wolfgang Köhler in *The Mentality of Apes* (1917). Her discussion of these experiments, as if from Sultan's point of view, allows Costello to demonstrate both the narrowness of "reason" as humans define it, and the importance of what she calls *sympathy*, the faculty "that allows us to share at times the being of another" (34). For Elizabeth Costello, "[t]here are no bounds to sympathetic imagination" (35), thus to the sympathy humans might, and should, share with animals. She makes this point again at the Faculty Club dinner that follows her first lecture, where fish is one menu choice, and where the strained conversation with her son, his wife, and the academics and administrators present, centers on the sacrifice, slaughter, and eating of animals.

Costello's second lecture, "The Poets and the Animals," delivered the following day in an English Department seminar room, turns to poetry (written by William Blake, Robinson Jeffers, Gary Synder, D. H. Lawrence, and Ted Hughes) that is itself "sympathetic," that attempts an "engagement" with, and an "embodying" of, animals. During the question period, topics range from ecology to cruelty and from hunting to vegetarianism. Her final session at Appleton is staged as a debate with professor of philosophy, Thomas O'Hearne, who presents three positions to which Costello is asked to respond: the first and second concern his reservations about the animal-rights movement and attribution of rights to animals; the third involves his contention that animals do not die and know death in the same way that humans do, and thus "to equate a butcher who slaughters a chicken with an executioner who kills a human being is a grave mistake" (64). Important in these second-day events are Costello's responses to O'Hearne and to those who question her after her lecture.

With the "ethics of sympathy" that Coetzee's Elizabeth Costello introduces to her largely indifferent academic audience, the issue is not whether "we have something in common—reason, self-consciousness, a soul—with other animals," but whether we can open ourselves to difference, to "the being of another" (34). Carol Adams in "'A Very Rare and Difficult Thing': Ecofeminism, Attention to Animal Suffering, and the Disappearance of the Subject," suggests that Western dualism, with its rationalist bias and concern for achieving objectivity, avoids "sympathy as a basis for ethical treatment" (594). Adams, whose work we refer to in an earlier chapter, characterizes an ethics of sympathy by "attention" to the suffering of nonhumans, for it is attention to suffering, she says, that "makes us ethically responsible"—not as atomistic, self-made individuals, but as related and interdependent selves able to ask of nonhuman animals, "What are you going through?" (601). Perhaps, as Jacques Derrida suggests in *The Animal That Therefore I Am*, with this question ("What are you going through?"), the word *can* (*pouvoir*) changes sense: "Henceforth it wavers" (Derrida 27), as if exposing the limits of human *can-have* capacity.

In *The Lives of Animals*, the "reflections" that supplement the fictionalized lectures respond in different ways to Coetzee's text. In their contributions, both Wendy Doniger and Barbara Smuts approach something like the "ethics of sympathy" espoused by Elizabeth Costello. Doniger addresses traditions of compassion for animals and vegetarianism in Hinduism, Buddhism, and Jainism, offering a more damning picture of animal sacrifice than that put forward by Kimberley Patton, and following Coetzee in "shifting the ground from the thoughts of animals to their feelings" (103). Barbara Smuts relates something of her experience dwelling among animals in the wild, particularly among baboons in Africa, where she learned "the rudiments of baboon propriety," and found herself "sharing the being of a baboon because other baboons were treating me like one" (110). Hers was the experience of learning about baboon individuality, of coming to know each one of the 140 animals in the troop as "a highly distinctive individual" with a particular gait, a face like no other, a characteristic voice, favorite food, favorite

friends, and favorite bad habits (111). Smuts parallels her narrative of living with baboons and gorillas with the story of her relationship with her rescue dog Safi, concluding by urging her reader "to open your heart to the animals around you and find out for yourself what it's like to befriend a nonhuman person" (120). Perhaps it is Marjorie Garber, however, who glimpses the crucial point behind Coetzee's book: she suggests that Costello's lectures are not so much about animals as about a debate between poetry and philosophy. Although the latter dominates and seems to win (79), Coetzee, she says, has raised the question whether poetry (literature) has something to offer animal studies. As she writes, citing from *The Lives of Animals*: "'Do you really believe, Mother, that poetry classes are going to close down the slaughterhouses?' asks Coetzee's John Bernard, and his mother answers, 'No.' 'Then why do it?' he persists. That is indeed the question" (84).

Peter Singer takes the question up in his response to Coetzee's book, a response in which, somewhat condescendingly, Singer, the philosopher, adopts a quasi-fictional form himself. His reflection relates the story of a discussion between a father (clearly Singer) and his daughter as to how equal consideration of interests might work in the case of a house fire that forced on the father the choice of saving either his daughter or the family dog. The father assures his daughter that, in such a situation, he would save her, for the reason that "*normal humans have capacities that far exceed those of nonhuman animals, and some of these capacities are morally significant in particular contexts*" (87; emphasis mine). Singer insists on calculation of capacities, rather than the "radical egalitarianism" (91) and rather than the reliance on "feelings" (89) that he reads in Coetzee's fiction. But it is the *fictional form* that is his main point of contention. As the father says to his daughter, "I prefer to keep truth and fiction clearly separate" (86)—*truth* being the purview of philosophy and not of literature. Moreover, "Coetzee's fictional device enables him to distance himself" from the arguments Elizabeth Costello presents. Because *The Lives of Animals* is fiction, Singer (the father figure) argues, "Costello can blithely criticize the use of reason, or the need to have any clear principles or proscriptions, without Coetzee really committing himself to these claims" (91).

Why literature? If we follow Peter Singer, we have to assume that no matter what *The Lives of Animals* says about human maltreatment of animals, fiction is not the place in which to deal with such things on an ethical level: as Plato established long ago in inaugurating the tradition of metaphysics, *logos* and *mythos* belong apart. Coetzee (because he resorts to fiction) is a "skeptic" when it comes to rationality as the sole ground of morality, Singer claims further in his Foreword to Paola Cavalieri's *The Death of the Animal*, the book we discuss in chapter 5. Or, to put the point in Cavalieri's own terms, Coetzee and his literary fictions do not belong to the realm of "narrow morality," which has to do with assessment of interests and with questions of right and wrong. Clearly, if Coetzee's work serves as an example, Peter Singer, "father" of animal studies and "father" in his fictionalized response to Coetzee, would have analytic philosophy define and contain the field; and if the field is to serve what he considers to be animal ethics, no more than continental philosophy would literature be taken seriously as contributing to it. Several other questions are at issue in Singer's "Reflection," among them: What is the proper mode of academic address—of an address such as Coetzee was invited to deliver at Princeton University?

In a markedly different view from that of Singer or Cavalieri, Carrie Rohman in *Stalking the Subject*, the book that we mention at the opening of this chapter and that we take here to represent a work of literature and literary criticism, suggests that the field of animal studies "emerges from the legacy of poststructuralism and its attendant analysis of subject-formation, at the same time that its interest in the radically 'other' pushes the recent 'turn to ethics' in literary studies beyond the familiar boundaries of the human" (9). In other words, critical animal studies, for her, is not a continuation of humanism so much as an interrogation, and revision, of its foundational self or subject in an attempt to move beyond the either/or boundaries this subject invariably puts in place. In both senses, the field of animal studies assumes the task of "illuminating the complexities of the subject's relation to animality within and beyond symbolic codes" (8)—a task that, as we note in earlier chapters, is not embraced by those analytic philosophers who have come to dominate the field. This task includes deconstruction of several dimensions of the metaphysical subject or self, among them: the links between subjectivity and speech (10; 17), and between subjectivity and carnivorous sacrifice (14); the tradition's parallel objectification of female body and animal body (15); the notion of sovereignty; the dialectic between speciesism and humanism (20); the dialectic between Darwinism and Freudianism (22); and Freud's suggestion that "an attempted rejection of humanity's own animality created the human unconscious" (23).[1]

Always with a view to the ontological place of the animal, and to questions of animal consciousness and language, Rohman turns her attention in this book to a number of modernist literary writings, for example: the poetry of T. S. Eliot, particularly "Sweeney Among the Nightingales," whose central figure, Sweeney, "is one of the clearest animalized humans to be found in the literature of modernism" (31); Joseph Conrad's *Heart of Darkness*, particularly as to "the pivotal nature of the animal" in this novel (41); D. H. Lawrence's *The Plumed Serpent*, which "provides a studied meditation on the discourse of animality and its relationship to consumption" (53); H. G. Wells' *The Island of Dr. Moreau*, which "asserts the coincidence of human and animal and suggests that the denial of one's animal nature cannot be sustained" (64); Wells' novel, *The Croquet Player*, for characterizing animality as a "haunting" and as a "contagion" that defies the powers of repression (64); and Djuna Barnes' *Nightwood*, a text that "engages us in deep philosophical narratives that trouble humanist subjectivities by privileging a kind of animal consciousness" (133). Rohman's book offers a sustained study of D. H. Lawrence, in whose work "the tension between destroying and acknowledging the radical alterity of the animal other recurs" and "reveals the deeply troubling relationship between human and animal in the modernist moment. Lawrence's work also signals a deepening disgust with humanism in its rationalist mode" (100).

Why literature? Rohman suggests in the conclusion to *Stalking the Subject* that her book explores representations of the subject (the human animal) in modernist literature, "but in a way that radically destabilizes the terms under which that subject has almost always been understood" (160). She takes studies such as hers, coming out of literary criticism, to have definite ethical import, although she hesitates to define "ethics" as do Peter Singer, Paola Cavalieri, and a number of other analytic philosophers. For as Rohman notes, "ethics itself, typically understood as involving rational subjects who can propositionally agree to the terms of a moral community, usually remains embedded within a proscriptive 'human' grid." Her study approaches ethics differently. *Stalking the Subject* "asks us to radically reconsider the terms of ethical inquiry, the contours of the 'other,' and the assumptions about subjectivity that are often profoundly embedded in our vision of what constitutes legitimate theoretical and scholarly work" (160).

10

TURNING POINTS

This book has attempted a broad survey of current contributions to critical animal studies, drawing on many, but by no means all, of the multiple, multidisciplinary resources that have been brought to this rapidly growing field. Addressed largely to undergraduate and graduate students and to readers who are simply interested in animal studies, the book provides a more-or-less academic introduction to the field, one that does not take account of the contributions made to animal studies, historically and currently, by activist individuals, institutions, and organizations—this as calling for an extensive study of its own. Over the course of the book, while broaching such issues as industrial farming, wildlife resource management, animal captivity and display, hunting, and experimentation, a number of particularly challenging problems have emerged pertaining to: the thinking of animal life and of the difference between nonhuman and human animal life; concepts of the human self or subject that are constitutive of hierarchical human/animal, same/different binaries; approaches to ethics that are rooted in a "like us" standard and that rely on preprogrammed calculations; adherence to a post-Cartesian reliance on mind or mental capacity, and so on. Proponents of specific philosophical or ethical approaches may find little agreement as to how critical animal studies might best proceed from here, but most return in one way or another to these same theoretical problems, either to defend a given position, or to subject it to critique.

We have suggested throughout the book that such critical analysis is essential to "critical" animal studies, and that probing legacies and prevailing binaries may well provide ways forward from here. By way of conclusion, the following discussion highlights seven theoretical areas or issues under scrutiny and debate in the field, seven challenging contexts out of which might come critical *turning* points—points on which, that is, critical animal studies might *veer*, to invoke Nicholas Royle's figure in *Veering: A Theory of Literature*. I like the word (which is at once literal and figurative) in that, as Royle explains at the outset, and then develops provocatively in subsequent pages of his book, veering "is kinetic and dynamic," offering "a mobile arsenal of images and ideas for thinking differently;" impelling us "towards new questions," and offering "fresh slants." Not the least, "[v]eering is not human, or not only human. Other animals veer. So do objects, such as stars. The theory of veering is non-anthropocentric. It gets away from the supposition that we human animals are at the centre of 'our' environment" (viii).

1. Ethics

In his 1789 treatise, *The Conflict of the Faculties*, Immanuel Kant provides an architecture for the university, a teaching institution that would be autonomous, hierarchical, and governed by the rational philosophy that, given its role in overseeing the other disciplines, he positions at the institution's uppermost point. *The Conflict*, a founding text, spells out what Richard Rand calls "the blueprint for the modern research university" (Rand 1992, vii): it led to Wilhelm von Humboldt's 1810 plan for the University of Berlin, and by the end of the nineteenth century it had been adopted by American universities as well. In many ways, today's Western research university remains a Kantian institution, a philosophical institution that defers to philosophy on matters requiring rational analysis, as much as to ethical discriminations between the true and the false, what is right and what is wrong. And, in its continuing role as the institution's overseer, philosophy still declares which disciplines do, and do not, belong within the boundaries of rational thought. Derrida suggests in his "*Mochlos*" essay that the Kantian University—modeled on the fantasy of an upright male body that is animated rationally from its top—has grown old, enfeebled, and in need of some critical work on its foundations. Insofar as animal studies remains *informed* by this institutional model, it too requires regrounding (McCance *Medusa*, 27–46).

We can see evidence of the problem in an institutional definition of ethics. In the Anglo-American tradition out of which animal studies is said to have emerged, "ethics" is understood as the sole purview of prescriptive philosophy. The latter has been challenged from outside its borders by literary scholars, architects, geographers, feminists, continental philosophers and others, who argue that such an ethics is too reliant on powers or capacities, particularly mental capacities, that it attributes to "normal" adult humans and denies to animals who are not sufficiently "like us." Such an ethics has been faulted as Cartesian, and given that its moral standard is based on a seventeenth-century ideal of *homo rationalis*, as narrow and androcentric. From within the tradition of analytic philosophy, however, ethicists have challenged their critics, not the least in ways that reinstate the *logos/mythos* binary, as Peter Singer does in distinguishing ethics, as the realm of "fact," from the "fiction" of J. M. Coetzee.

"'The animals! They're two-thirds dead. Do you not understand that?'" The question is posed in Yann Martel's latest novel, *Beatrice & Virgil*, directed to the central character, Henry L'Hote, by a taxidermist, who also calls himself Henry. It may seem odd to have a taxidermist, who makes a living from dead animals, lamenting the death of animals, particularly when he utters this question while inside his shop, Okapi Taxidermy, surrounded by dead animals of all sizes and shapes. "In quantity and variety, put together, two-thirds of all animals have been exterminated, wiped out forever," the taxidermist goes on (134–135). The Henry to whom he addresses his concern is, like Yann Martel, a novelist whose work has met considerable fame—Martel's *Life of Pi*, won the prestigious Booker Prize in 2002. During the course of the novel, Henry's wife gives birth to a son, Theo, which happens also to be the name of Martel's infant son. There are additional parallels between the novel's novelist Henry L'Hote and the book's author Yann Martel, as between Martel's novel and Coetzee's *The Lives of Animals*. For example, just as Coetzee's Tanner lectures embed fictional lectures, so Martel's novel embeds another novel, *Beatrice and Virgil*, which Henry L'Hote begins writing just as Martel's novel ends. Both texts eschew so-called rational discourse in favor of fiction and a willful blurring of philosophy's and traditional autobiography's truth/fiction dichotomy. We can read these texts, then, not only as raising sensitive and pressing animal ethics issues (both compare today's extermination of animals to the Holocaust), but also, because they do so from the underside of Enlightenment philosophy's *logos/mythos* binary, as calling the very meaning of "ethics" into question. Critical animal studies has opened a much-needed reevaluation of the truth/fiction hierarchy that belongs centrally to metaphysics, and that equates philosophy with a first-person subject's truth-telling exposition, or as Mieke Bal calls it, *exposé*.

2. Anthropomorphism

Scholars surmise that even in prehistory, humans endowed animals with anthropomorphic features, sometimes as either deities or demons. We are ignorant of many of the meanings that such anthropomorphized animals had in prehistoric and ancient human imagination, although a book like Jorge Luis Borges' *The Book of Imaginary Beings* is replete with possibilities, not all of them benign. Of the Celestial Stag, for example, Borges writes that these tragic animals, belonging to Chinese mythology, are said to live underground in mines and to desire nothing more than to reach the light of day; hence, they implore miners to lead them to the surface, bribing them with the promise of revealing veins of silver and gold, and when this gambit fails and they become troublesome, the miners must wall them up in shaft—or risk being tortured to death by them (55). These days, partly as a result of critical animal studies, we are less comfortable—if at all comfortable—with anthropomorphism, the attribution to animals of human characteristics, feelings, or motives, as is commonplace in fairy tales, children's stories, films, and cartoons. For studies such as Steve Baker's *Picturing the Beast: Animals, Identity and Representation* demonstrate, even Rupert Bear stories can be coded for racial and sexual difference (126–138), and self-centered speciesism would seem to lurk behind anthropomorphism in any case.

In *Thinking with Animals: New Perspectives on Anthropomorphism*, historians of science Lorraine Datson and Gregg Mitman, acknowledge that, "[c]onsidered from a moral standpoint, anthropomorphism sometimes seems dangerously allied to anthropocentrism: humans project their own thoughts and feelings onto other animal species because they egotistically believe themselves to be the center of the universe" (4). And even "if anthropomorphism is decoupled from anthropocentrism, the former can still be criticized as arrogant and unimaginative" (4). Characterizing the behavior of a herd of elephants as akin to that of a large, middle-class American family, or dressing up a pet terrier in a tutu, strikes critics of anthropomorphism "as a kind of species provincialism, an almost pathological failure to register the wondrous variety of the natural world"

(4). Another moral dimension of anthropomorphism that Datson and Mitman deplore is one we have encountered in this book as prevalent in animal studies: arguments that base the moral worth or rights of animals on their likeness to humans where certain capacities are concerned (4–5). Having seen several examples of such anthropomorphism in earlier chapters of this book, we might appreciate that this is a context in which much critical work needs to be done. Many contributors to animal ethics deplore the "like us" standard. Do the years-long efforts of researcher H. Lyn White Miles to "civilize" an orangutan, Chantek, comprise but another version of this standard? In her contribution to *The Great Ape Project*, Miles explains that, under her tutelage, Chantek was toilet trained, taught to eat with utensils, "distracted from masturbating in public, and not allowed to peek under the toilet doors at others using the facilities" (52). Most importantly, Chantek was tutored, with some success, to communicate using sign language, a result that, for Miles, does not challenge the traditional oppositional human/animal paradigm, so much as move the orangutan to the human side of the line. "Chantek has met the Cartesian definition of personhood," she writes, "at least at the level of a young human child" (51). It was Descartes who declared the importance of "real speech" in establishing human/animal difference. In his contribution to *The Great Ape Project* volume, Harlan B. Miller seems quite in line with the Cartesian standard: apes, such as Chantek, he suggests, merit the status of "quasipersons," a standing "different from, and generally lower than, that of fully fledged persons" for the reason that quasipersons lack speech (230–236). Still measuring apes against the traditional "like us" personhood standard, Robert W. Mitchell in his essay "Humans, Nonhumans and Personhood" in the same volume, concludes on a similar note: apes, lacking relative to humans in self-awareness and linguistic skills, are "not fully persons," although still "deserving of our protection" (244–245).

What interests Datson and Mitman, however, is not the anthropomorphic and/or anthropocentric debate over "who counts" as having moral worth close to, or equal to, humans. Their question rather is: "Can we really ever think *with* animals?" (5). The "with" here is important. Their question is not how humans think *about* animals but whether, and how, they can think sympathetically *with* them (5). "In certain historical and cultural contexts," they write, "the longing to think with animals becomes the opposite of the arrogant egotism decried by critics of anthropomorphism" (7). Thinking with animals, Datson and Mitman suggest, is an impossible, but genuine, attempt at "complete empathy," a relinquishing of self in favor of the other (7). This recalls the "ethics of sympathy" advanced by some literary and feminist contributors to animal studies, who argue that sympathy and compassion extend ethics beyond the narrow confines of "like us" approaches given to calculating capacities.

Whereas the Datson and Mitman anthology centers on the anthropomorphizing of animals, Mark S. Roberts in *The Mark of the Beast: Animality and Human Oppression*, deals with the "reverse" ("zoomorphic") practice of attributing animal features and status to humans, and by extension, to animals, too, the latter cast as the "beasts" in Roberts' title. We consider some zoomorphisms in chapter 6 of this book, critical practices through which the human/animal division is blurred. Indeed, in the discourse of animal studies, animal figures abound, some descriptive of critical practices themselves as involving, for example, trekking, pursuit, or the tracking of a scent. As with anthropomorphism, such "zoo" practices can cut either way. The practice of casting humans as animals has worked historically in pernicious ways, enabling the domination, even slaughter, of certain classes or groups of humans, something Roberts demonstrates in *The Mark of the Beast*—as giving good reason to be critically attentive to how traditional theories, modes, and acts of what he calls "animalization" are interpreted and restored in the discursive practices and actions of contemporary thinking (x–xi).

3. Dualism

In his *Description of the Human Body and All of Its Functions*, an unfinished treatise dating to the winter of 1647–1648, Descartes remarks that, "If you slice off the pointed end of the heart in a live dog, and insert a finger into one of the cavities, you will feel unmistakably that every time the heart gets shorter it presses the finger, and every time it gets longer it stops pressing it" (317). As we note in earlier chapters, it was by way of his experiments in dissection and vivisection that Descartes cut Western modernity's mind/body, life/death, man/animal lines of division with the precision of an anatomist's knife, accounting for the animal body as nothing but a machine, set apart in the human from a wholly immaterial mind. As part of what Edmund Husserl called this "epoch-making" turn that philosophy takes in the Cartesian text, the inward turn "toward the subject himself" (4; 2), Descartes identified speech with thought, for example, writing to the Marquess of Newcastle to explain that "none of our external actions can show anyone who examines them that our body is not just a self-moving machine but contains a soul with thoughts, with the exception of spoken words" (303). For Descartes, it was crucial that such words, utterances of "real speech" that manifest the thinking part of the soul, not be confused "with the natural movements which express passions and which can be imitated by machines as well as by animals" (*Discourse*, 140). Here is as good a statement as any of what Cartesian dualism entails. And because animal studies has not yet moved beyond dualism in its various forms, including the valorization of speech in its proximity with mind or mental capacity, this is a theoretical and practical issue that calls for critical questioning.

Such work has begun, and we have surveyed some of it in this book, albeit without sufficient focus on *speech*—or the mouth, the organ of its production. In "Unum Quid," French philosopher Jean-Luc Nancy dwells on this organ and in the process blurs the dualism that Descartes attempted to put into place through it. As Nancy points out, the mouth actually functions in Descartes' work as a "common place," a sort of "joint" between soul and body—a *sort of* joint, the "quasi" *permixtio* that Descartes relies on in his Sixth Meditation, where he ventures that, rather than *entirely* separate and distinct, soul and body are intermingled to form a certain unity, a "quasi-single whole" ("Unum," 132–133; 160–161). Thus, in Nancy's rereading of Descartes, Derrida notes in *On Touching: Jean-Luc Nancy*, the mouth is at once a mechanical instrument, an organ of speech production, and what opens to (opens into *exteriority*) "the soul of the one whose being it is to utter" (29). The mouth is the "common place of the incommensurables in question" (25), incommensurables that include soul (speech, thought) and body, interiority and exteriority, animate and inanimate, and not the least, human and animal. The mouth opens to something like a whole, rather than to the open/closed opposition that animal studies still takes from Descartes, and it suggests that, even for him, Cartesian dualism could not be sustained. Might such critical rereadings of tradition such as Nancy undertakes contribute to an ethics of "sharing out without fusion," an ethics that is distinguished both by "participation *and* partition" (Derrida *Touching*, 199; 195)?[1]

4. Rights

As defined in *Black's Law Dictionary*, a right is a "power, privilege, faculty, or demand inherent in one person and incident upon another. Rights are defined generally as powers of free action" (1486). By and large, moral theory has adopted the political-legal definition, increasingly in the 1970s and 1980s with the development of new reproductive and biomedical technologies. Thus, Tom Beauchamp and James Childress write in *Principles of Biomedical Ethics* that, "Most recent writers in ethics recognize that 'rights' should be defined in terms of claims. In our framework, rights are best seen as justified claims that individuals and groups can make upon others or upon society" (48). Similarly, the individualism and adversarial nature of rights has come to the forefront in animal ethics, for example in the works of Tom Regan that we consider in this book. Many argue, however, that in a world of shrinking resources and profound imbalances between those who "have" and those who "have not," the understanding of rights as *absolute* claims in conflict makes little social or ethical sense.

On the one hand, in "Violence Against Animals," his dialogue with Elizabeth Roudinesco, Jacques Derrida lauds the tremendous progress being made today in primatology, with its discovery in the "higher primates" of "extremely refined forms of symbolic organization: work of mourning and of burial, family structures, avoidance if not prohibition of incest, etc." (66). On the other hand, he suggests that extending the concept of human rights to animals "leads to naïve positions that one can sympathize with but that are untenable" (64). Why? "A certain concept of the human subject, of post-Cartesian human subjectivity, is for the moment at the foundation of the concept of human rights" (64–65). Thus, "to confer or to recognize rights for 'animals' is a surreptitious or implicit way of confirming a certain interpretation of the human subject, which itself will have been the very lever of the worst violence carried out against nonhuman living beings" (65). As Kelly Oliver puts it, encapsulating the argument of many contributors to animal studies, "[r]ights may be better than nothing, but they still leave oppressive power structures intact" (*Lessons*, 31).

5. Machine

As we note in an earlier chapter, architectural theorist Catherine Ingraham refers to today's world as "post-animal," a world in which humans have left their animalness behind, and thus turned their backs on animals (*Architecture* 16). David Wills recognizes this in *Dorsality: Thinking Back Through Technology and Politics*, where he contends on many levels with what he calls humankind's "dorsal" turn. In one sense, dorsality implies the moment in which, assuming upright posture, mankind "turns its back in a radical way on whatever is behind it. We know how it abandons the animal, refines the senses by downgrading smell and hearing, and reconfigures the knowable other within a frontal visual perspective" (8–9). With this dorsal turn, the human gradually "forgets" its animality in favor of an integral, forward-looking image of itself, one that understands language, along with the technologies of tool use, as being ahead, within its frontal field of vision, on the horizon of a forward progression. We have glimpsed the contours of this self-image in the humanist discourse of animal studies. It is a self-image that Wills confounds, however, by tracing the dorsal turn back, before assumption of upright posture and freeing of the hands to use tools, to the first "movement of a limb that, as the Latin teaches us, is the sense of *articulation*. Within that logic, there is technology as soon as there are limbs, as soon as there is bending of those limbs, as soon as there is any articulation at all. As soon as there is articulation, the human has rounded the technological bend" (3).

Indeed, well before the emergence of a limb, when "the animate first articulates and so becomes technological in the self-division of a cell," the dorsal turn has already occurred (3–4), which is to say that we cannot find an originary moment when humankind was not already inhabited from the inside by an "unassimilable otherness," an *inanimation* (12) that Western metaphysics invariably attempts to cast outside of itself and control. Dorsality, in this sense, names the "turn into a technology that was always there" (3), a "turning to see the technology of the human itself, inside itself" (7). It suggests a needed turning back to "an originary mechanics at work in the evolution of the species" (5), a turn that "would appear to contradict the definition of the technological as production or creation, as fabrication produced by hands manipulating matter within a visible field" (7), and that would operate as a form of "resistance precisely to a technology that defines itself as straightforward, as straight and forward, straight-ahead linear advance" (6).

Perhaps, what Wills calls *dorsality* might be a turn, a *turning* (*veering*) that critical animal ethics itself needs to make, a radical rethinking of life and of the animal life on which the human has turned its back, a rethinking as well of "the fact of a relation between *bios* and *techne* so complex and so historic that any presumption of the priority of one over

the other can be sustained only by means of an appeal to the metaphysics of creation" (5).[2] For Wills, such rethinking would necessarily involve
the rereading of what Sigmund Freud called the *death-drive*, a "sort of
primary autodeictic and autobiographical movement or moment of life"
that renders impossible any easy bifurcation of animate from inanimate,
bios from *techne*, or human from animal (see Wills "Order," 37).

6. Passivity

In *Simians, Cyborgs and Women*, we have noted, Donna Haraway questions the semantics of invasion and defense that typically characterizes
immune system discourse and that casts the self as a defended stronghold,
assured by its immune system of invulnerability and self-integrity over
against a hostile other-outside. From her point of view, such discourse
misrepresents life, which "is a window of vulnerability. It seems a mistake to close it" (224). David Wills makes a similar point in *Dorsality* and
takes it further, relating vulnerability to the meaning of ethics: "By means
of dorsality, sexuality and human relations in general are marked by an
extreme vulnerability, by the sort of passive trust that would be a condition of possibility for ethics in general" (13). For Wills, "one can begin
to be ethical only by respecting the most vulnerable forms of life" (13).
Perhaps Derrida, in raising the issue of vulnerability himself, takes the
point even further still, suggesting that it is only *from the other* that vulnerability—finitude, the relation to death—is made available to the self,
to the self who would thus be enabled to partake in ethics. In his Introduction to *Philosophy & Animal Life*, a book that, among other things,
asks what vulnerability has to do with our understanding of the human
and human-animal difference, ethics, justice, and philosophy, Cary Wolfe
considers the meaning that vulnerability has for Derrida, as well as commenting on Cora Diamond's reading of Derrida on this point.

Diamond's "The Difficulty of Reality and the Difficulty of Philosophy" (in *Philosophy & Animal Life*) picks up on vulnerability by way of a
poem written by Ted Hughes, *Six Young Men*, in which the speaker looks
at a photo from 1914, the beginning of World War I, of six smiling young
men—who were all dead within six months of the photo's having been
taken. What interests Diamond in the poem, she says, is its expression

of a kind of limit on the human faces, "the experience of the mind's not being able to encompass something which it encounters" (44). This is the kind of experience she reads in Coetzee's *The Lives of Animals*, in the character Elizabeth Costello, whose lectures present "a kind of woundedness or hauntedness, a terrible rawness of nerves. What wounds this woman, what haunts her mind, is what we do to animals" (47). The story of Costello's lectures is about the "presenting of a wounded woman" (49), Diamond says, rather than about presenting arguments, as the responses printed in *The Lives of Animals* suggest. Elizabeth Costello does not engage in arguments and does not take seriously the conventions of argument (52). Rather, through her, "Coetzee gives us a view of a profound disturbance of soul" (56). Borrowing Stanley Cavell's word "exposure" (which turns what Bal calls *exposé* on its head), Diamond characterizes our position vis-à-vis animals as that of being exposed to "the bodily sense of vulnerability to death, sheer animal vulnerability, the vulnerability we share with them" (74).

The word *exposure* might recall Derrida's sense of vulnerability before his little cat, looking at him naked. As we note earlier, at the opening of *The Animal That Therefore I Am*, Derrida recounts a ritual early-morning scene in which, on his way to the bathroom, he is caught in the gaze of his cat, exposed, naked before his cat—a situation not of seeing but of being seen, and of becoming aware of his own nudity, his vulnerability, before the gaze of the other. *Vulnerability*: for despite the fact that the animal in question is Derrida's own little cat, behind its gaze, "there remains a bottomlessness," an alterity that, even in the case of one's own pet, is finally "uninterpretable, unreadable, undecidable, abyssal and secret" (12). Derrida relates his experience of a profound sense of powerlessness when brought up against the "unsubstitutable singularity" of this animal, this "existence that refuses to be conceptualized" (9), to that of being stripped bare of the mantle of autonomous first-person author, and of being called on as an addressee, as one who comes after or "follows." In the end, according to Derrida, ethics finds itself in this position of being-seen and of not-knowing, of not having the power, the *capacity*, to name or define "the animal." For both Derrida and Diamond, as Cary Wolfe points out, this vulnerability, this awareness of *not-having*, is the very possibility of ethics: that is, of an ethics concerned, not with calculability or capacity, but with justice (18). This is critical terrain, around which animal studies cannot detour.[3]

7. Sacrifice

In this book, animal sacrifice has been a recurrent theme, one we encountered, for example, in Mieke Bal's analysis of dog sacrifice as part of museum *exposé*; in Elizabeth Costello's references, in her fictional lectures at Appleton College, to the slaughter and eating of animals; in Wendy Doniger's response to Coetzee's *The Lives of Animals*, with particular recourse to the Hindu tradition of blood sacrifice; in Kimberley Patton's plea for understanding of the structure and role of ritual sacrifice in certain world religions; in Carrie Rohman's suggestion of a link between theories of subjectivity and carnivorous sacrifice, and so on. We have also considered sacrifice as involving what, in the Preface, I refer to as a "nonliteral putting to death." David Clark addresses the latter in "On Being 'the Last Kantian in Nazi Germany': Dwelling with Animals After Levinas," where he points out that, according to Derrida, animal sacrifice is "symptomatic of a generalized carnivorous violence, a 'carnophallogocentrism' modeled upon the 'virile strength of the adult male'" (176). For Derrida, Clark remarks, "the point is not that we must stop eating meat—as he says, the distinction between animal and plant 'flesh' is itself suspect—but to think critically about how carno-phallogocentric discourses and regimes" (177) perpetuate domination and assimilation of the other. In other words, Derrida's writing on sacrifice is one instance where "the question of the animal" opens to many of the "great questions" of our day, several of which we touch on in this book and even in this concluding chapter, particularly that of the "subject," the autonomous author of ethics, and as a "normal adult male," the standard against which the moral worth of others is measured. What this subject does not recognize as sufficiently the same as a universalized "us," is sacrificed as inadmissible—a schema, a *sacrificial structure*, that, Derrida ventures "is still today the order of the political, the State, right, or morality" ("'Eating'" 114).

Carnophallogocentrism combines *carno* (of flesh, flesh-eating, or sacrifice), *phallo* (from the Greek *phallos* and Latin *phallus*, a penis; hence "masculine"), and *logos* (word, speech). The term suggests an essentialist framework that is patriarchal, privileging the "virile strength of the adult male, the father, husband, or brother" ("'Eating'" 114); that entails a literal and figurative "eating" of the other (assimilating of difference), justified by way of same/different, self/other, man/animal binaries; and that gives priority to speech, and to the proximity between speech and thought, mental capacity, or mind. The term, then, has everything to do with both the symbolic and real *mouth*, with an "eat-speak-interiorize" (114) way of relating to the other. In a carnophallogocentric schema, in short, "the metonymy of 'eating well' (*bien manger*) would always be the rule" (114).

As a graduate student who abandoned animal experimentation, I had not yet learned to think critically about what Derrida calls the *metonymy* of "eating well," this nonliteral, or at least "noncriminal," putting to death. Hopefully, having worked your way through this book, you have become a more critical reader of the "eat-speak-interiorize" sacrificial structure, and now are ready to answer the question with which the book opens. As Derrida suggests in the interview "'Eating Well,' or the Calculation of the Subject," this question, along with everything it opens and calls us to rethink, is one to which animal studies today must respond—not so much with a ready answer, quick in coming, as with meticulous critical analysis of what, historically, in theory and in practice, connects the "subject," and the subject of "ethics," to "sacrifice."

Notes

Chapter 1

1. Immanuel Kant exemplifies this subject-centered gesture in his *Anthropology*, where he declares that, "[t]he fact that the human being can have the 'I' in his representations raises him infinitely above all other living beings on earth. Because of this he is a *person*, and by virtue of the unity of consciousness through all changes that happen to him, one and the same person—i.e., through rank and dignity an entirely different being from *things*, such as irrational animals, with which one can do as one likes" (15).

Chapter 2

1. Citing American statistics, James H. Shideler notes in "The Farm in American History," that the 1990 U.S. census showed scarcely 2% of the 4.9 million of the population to be farm people, a monumental shift from the first census in 1790, where 93% of Americans were reported as farmers (18).
2. Joyce D'Silva explains that: "Pigs are 'switched on' to their highly sensitive snouts, which, in natural conditions they would use for exploring their environment and for rooting in the soil for roots and grub to eat. With those two choices removed, rooting can transform into biting all too easily and young pigs, who love to indulge in play and dashing about, can be seen fighting and biting each other instead" (37).

3. In the Afterword to the 2007 edition of *Slaughterhouse*, Eisnitz points out that as recently as 2005, the United States Department of Agriculture, in "flagrant disdain" for the Humane Slaughter Act, decided to classify rabbits as poultry. The reason: "Species that the USDA deems to be 'poultry' (including the nine billion chickens and turkeys slaughtered each year) are excluded from the Humane Slaughter Act" (310). Eisnitz goes on to document some atrocities that are involved in industrialized rabbit slaughter.

4. Studies on the connections between animal abuse and human violence, for example, family violence and abuse of children, have begun to emerge in animal studies. (See, for example, Andrew Linzey's edited collection, *The Link Between Animal Abuse and Human Violence*.) Surely, the slaughterhouse is an environment conducive to such research.

5. Partly as a result of industrialized agriculture, climate change is happening much faster than even the most pessimistic models have previously suggested, with the Arctic warming at twice the rate of the rest of the planet. Canadian researcher David Barber, a sea ice specialist and lead investigator of the Circumpolar Flaw Lead System study involving some 300 scientists from around the world, compares the melting of sea ice in the Arctic to the disappearing rain forests: just as cutting down all the trees in the rain forest will lead to collapse of that ecosystem, so removal of sea ice will wipe out species, contribute to disease spread, and take away the livelihood of people living in the North. For more information on this, see the ArcticNet website: www.arcticnet.ulaval.ca.

6. "Under the skin of the big Arctic predator: Focus on environmental pollutants, bone system and reproductive organs in the East Greenland polar bear," an as-yet unpublished paper presented by wildlife veterinarian and toxicologist Christian Sonne at a May 2011 conference on Arctic climate change and pollution at Copenhagen University, documents this polar bear evidence.

7. Asking why the spread of "mad cow" disease in the mid-1990s occasioned such panic in Britain and around the world, postcolonial theorist Helen Tiffin points first to evidence that BSE could be transferred from cattle to humans through the consumption of flesh, including *cooked* meat. Such crossing of species boundaries challenged the human/animal binary, she suggests. Equally disturbing was the information that factory farm cattle, no longer the

ruminant herbivores of the era of family farms, had been turned into carnivores, whose food included processed parts of other animals. Humans have generally refrained from eating carnivores, Tiffin says, because the practice smacks of cannibalism—and because cannibalism, attributed to "savages," has been a limit marker for the civilization/savage binary (11–19).

8. Tzachi Zamir's *Ethics and the Beast* counters Singer's anti-speciesist position, arguing that many forms of speciesism are consistent with animal liberation.

9. That the beleaguered stone has not fared well in Western philosophy owes to the tradition's animate/inanimate dichotomy, a binary that Singer calls on in this passage. For a provocative countering of this binary, see *Chora L Works*, which documents correspondence and design theory relating to Derrida's collaboration with architect Peter Eisenman on a "garden" that was to have been constructed in Bernard Tschumi's Parc de la Villette in Paris—not incidentally, on a site previously occupied by a slaughterhouse. The garden, in the shape of a lyre, was to have been made entirely of stone.

10. Marc Bekoff and Jessica Pierce note that researchers working with mice reported distress and a heightened reaction to pain in animals who observed their cagemates suffering and writhing in pain after being injected with acetic acid. This may not suggest "anticipatory dread" (86), but it does challenge some assumptions about animal indifference to impending suffering.

11. Singer supports abortion, infanticide, and euthanasia in certain situations and according to the arguments he presents in *Practical Ethics*.

Chapter 3

1. The word "abolition" recalls the movement to abolish American slavery.

2. Although objections to captive breeding programs have come from those who, in Goodall's words, "feel last-minute solutions will not work, and are a waste of time and money," she considers it fortunate that the biologists who worked to save the endangered or extinct-in-the-wild species mentioned here "refused to listen to them" (*Hope*, 5).

3. Meine writes that, after having defined the technical foundations of the wildlife resource management field, Leopold challenged his

professional progeny not to neglect a more complex task: "'I daresay few wildlife managers have any intent or desire to contribute to art and literature, yet the ecological dramas which we must discover if we are to manage wildlife are inferior only to the human drama as the subject matter for the fine arts.' Even as wildlife ecology was beginning to grow confident in its expanded role as a science, its chief scientist was advising its adherents to surmount 'the senseless barrier between science and art'" (Meine, "Moving" 19).

4. The "like us" standard that is so prevalent in the animal studies movement, and that has been shaped by Singer and Regan, takes some unusual forms: for example, in *Wild Justice*, Marc Bekoff, a cognitive psychologist, and Jessica Pierce, a philosopher, argue that morality, rather than peculiar to humans, is an evolved trait, and that other animals "have it just like we have it" (xi). (See also, Dale Peterson, *The Moral Lives of Animals*.)

5. Regan's 2004 Preface to *The Case for Animal Rights* considers and dismisses the criticisms of Deborah Slicer, Josephine Donovan, and other ethics-of-care feminists as to the patriarchal, hierarchical, rationalist, and dualist dimensions of his rights view. For more on this, see chapter 7 in this book.

6. Regan's wording in the passage cited—treating nonsubjects-of-a-life *as if* they have basic moral rights, even if that amounts to "giving them more than is their due"—is unfortunate. It suggests that moral rights belong "naturally" to humans, that they are "ours" to give: those belonging to the group *us*, represented by a rational adult human (moral philosopher), give rights to *them* who are "most like us." Yet this wording goes right to the heart of his understanding of ethics, a point we return to several times in this book.

7. As Marianne Dekoven points out, the "term *nonhuman* itself is ideologically loaded: only from the point of view of the human are other animals nonhuman. Donna Haraway uses 'critters' to designate all animals including humans (*When Species Meet*). *Nonhuman* is not an inevitable descriptor. There are many other ways to designate the beings in question: by species, habitat, ecological niche, relation to predation (predator or prey), type of nourishment (carnivore, herbivore, omnivore). *Non-white, non-European*, and *non-Western* are parallel to *nonhuman* and reveal what is at stake in using it. Indeed, may critics and theorists who work in animal and animality studies are motivated

by the parallels between animals and subjugated humans" (363).

8. On the conjunction of animality and disability, we might well wonder why the question of the being of the bee has been taken up by one philosopher after another in the Western tradition, from Aristotle through Schelling, Marx, Heidegger, and Lacan, a tradition that Jacques Derrida describes as given to thinking "the sense and senses of the bee, and the bee's reason for being" (Derrida and Roudinesco "Violence," 131), always in the interests of setting apart, in binary fashion, the being of the bee from the being of man. Thus, as Heidegger has it, the bee, *without being able to grasp* things *as such*, is "poor in world," and, unlike "world-forming man," cannot open itself in a questioning way to blossom, scent, or hive (241–249). Bee being belongs to what Heidegger calls "rigid fixity" (248), a theme that Jacques Lacan picks up in arguing that the bee's "system of signaling" in the wagging dance is not akin to human language, but only a "fixed code" (84–85). The speechless bee, deprived of language, is, to use Heidegger's word, *benommen*: dazed, stupefied, *dumb*. Aristotle, in the first lines of his *Metaphysics*, explains that the bee is dumb because it is *deaf*: bees are incapable of hearing sound, Aristotle claims, thus they cannot learn or be taught (1552).

9. See Regan's discussion in *Empty Cages* of fur mill fur, trapping in the wild, and seal hunting. As he explains in this text, it may be "going too far" to attribute minds and rights to fish, which is why he focuses in this book on the less controversial mammals and birds.

10. Insofar as geographers, like wildlife resource managers, propose reform measures rather than abolition, Regan would take issue with them as well. For example, situating themselves within the "new geography" movement, Chris Philo and Chris Wilbert call on humans "to desist fixing animals rigidly into our spatial orderings," and instead "to allow animals more space" (25). Grant animals more room, they say, even when they are traveling to the slaughterhouse in trucks; furnish them with more ecological resources *within* human settlements, "as in creating and leaving corridors of appropriate habitat threading through cities, towns, and villages" (25); and in general, "open up spaces wherein they can indeed exist and ignore us for most of the time, and which they can occupy and convert into their own beastly places, as many animals are continually seeking to do (even in the bustling city)" (25).

Chapter 4

1. Anita Guerrini, commenting on Aristotle's cruelty to animals, writes that: "The vascular system, for example, was difficult to discover in dead animals because the vessels collapsed after death. In living animals the veins were usually not visible to an external observer. The best solution, according to Aristotle, was to starve an animal and then strangle it. Looking at old men had taught him that blood vessels stand out more clearly in an emaciated body, and strangling prevented blood loss" (10).

2. Bernard Rollin maintains that even Aristotle implicitly recognized something like a husbandry contract when he affirmed that domestic animals benefit through their natural role of serving man (10).

3. Lori Gruen, citing Gross, alleges that Galen tied down squealing pigs, cut them open, and then severed their laryngeal nerves; even after the nerve was severed, the pigs continued to struggle (108).

4. Andrea Carlino points out that the *Fabrica* "marked a turning point in the history of anatomy," not only because of the revisions it made to human morphology, but also for reason of its success as a publishing venture. Indeed, she refers to the book as one of the most important and astute successes of the first century of printing (39). As is implicit in the "self-portrait" of Vesalius included in the book, the *Fabrica* established the importance of both the scalpel and the pen to the shift into modernity, even as it established the authority of Vesalius as both anatomist and author, one who speaks on his own authority. For a more detailed analysis of the role played by Vesalius in initiating the transition to modernity and its human/animal dichotomy, see McCance, "Anatomy as Speech Act."

5. I am using C. D. O'Malley's translation of the *Fabrica*, found in O'Malley's *Andreas Vesalius of Brussels 1514–1564*.

6. Guerrini notes that the use of monkeys and apes in research and experimentation has become increasingly controversial, partly because of the possibility of monkey diseases transferring to humans and partly because "studies of primate intelligence and behavior have revealed close similarities to humans" (4–5).

7. Noting that these reproductive technologies are now ubiquitous in contemporary agriculture, Twine writes: "The US company Bull's Eye Genetics provides a good example of the sort of technological infrastructure developed around AI [artificial insemination] in

cattle—selling bull's semen for 41 cattle breeds but also equipment such as semen tanks, thawing units and AI 'guns.' Such AI reproductive technology is widespread, especially in the cattle and pig industries. They already provide an important technological path—a pre-existing infrastructure for the dissemination of anticipated future prized genotypes" (95).

Chapter 5

1. In the roundtable discussion that follows the dialogue in *The Death of the Animal*, Cary Wolfe remarks that, "it seems odd, to me at least, that a philosophical orientation such as [Cavalieri's] would be presented in the form of a *dialogue*—that oldest of philosophical (or is it literary?) forms that would seem to unsettle the boundary between philosophy and literature in ways whose implications are not to be underestimated" (47). In other words, as Wolfe notes, dialogue itself is a literary form, one replete with rhetorical devices and dependent on two fictional characters, the interlocutors, in Cavalieri's case, Alexandra and Theo. In the end, then, it does not serve Cavalieri well as a form that would discriminate, hierarchically, between philosophy and literature.

2. As a follow-up to the above note on the *logos/mythos* binary, it is interesting and significant that, despite relying on this binary, Cavalieri, by availing herself of various "linguistic and literary means of persuasion," actually demonstrates what Cary Wolfe, following Derrida, refers to as the "intricated, interfolded relationship between philosophy and literature" ("Humanist," 48).

3. The "Roundtable" comprises responses to the Cavalieri's dialogue by professor of literature Cary Wolfe (see chapter 6), philosophers Harlan B. Miller and Matthew Calarco (see chapter 6), and novelist and literary critic John M. Coetzee (see chapter 9).

4. *The Beast and the Sovereign*, the two volumes of which have appeared in English translation, was the last seminar Derrida gave at the École des hautes études en sciences sociales in Paris. Volume 1 consists of thirteen *The Beast and the Sovereign* seminar sessions, dating from December 12, 2001, to March 27, 2002 (editorial note to Volume 1, xiii–xiv). The seminar deals with the history of sovereignty as intertwined "with that of a thinking of the living being (the biological

and the zoological), and more precisely, with the treatment of so-called animal life in all its registers (hunting and domestication, political history of zoological parks and gardens, breeding, industrial and experimental exploitation of the living animal, figures of bestiality and bêtise, etc." (editorial note to Volume 1, xiii). "On the permanent horizon of our work," Derrida writes in Volume 1, "were general questions about force and right, right and justice, of what is 'proper to mankind,' and the philosophical interpretation of the limits between what is called man and what is improperly and in the generic singular called the animal" (xiii–xiv). Volume II analyzes the history of the concept of *sovereignty*, including its association and dissociation of the figures of the *sovereign* and the *beast*, both, in different ways, "outlaws" (editorial note to Volume II, xiv). This volume brings together Daniel Defoe's *Robinson Crusoe* with Martin Heidegger's 1929/30 seminar *The Fundamental Concepts of Metaphysics* (xiv).

5. Underlying the modern concept of rights is a construction of the self or the subject that *gives itself* the authority and autonomy—the *power*—to name "the animal" in this reductionist way. This subject, historically and currently, is at the crux of hierarchical thinking. Derrida's work comprises a prolonged analysis of the structure of this auto-presentation: the self that *gives itself to itself* as autonomous reason, fully present to itself and therefore certain, clear, capable, self-determining, a bounded individual whose statements as to true and false are proffered as universal. This is the fantasy or "phantasm" of the self that Derrida refers to in *Rogues* as *ipsocentric*; the self of *ipseity*, where the word *ipseity* suggests "some 'I can,' or at the very least the power that gives itself its own law" (11). The sovereign "I can" is uncompromised insofar as the self believes that it can exclude from its autonomous sphere a whole differentiated field of experience, all traces of descriptive and/or empirical knowledge—and all trace differences that divide self-presence in favor of an essential noncoincidence. Here is one point on which Derrida's analysis of "the animal" differs radically from Cavalieri's: his contention that the concept of the self undergirding Western metaphysics—the self giving itself to itself in an unbreached immediacy, holding itself up as standard and norm—always a phantasm, has now become untenable. And with the loss of faith in that metaphysical phantasm goes "the animal" over which the ipsocentric self has for centuries presumed to have sovereign power.

6. In "White Mythology," as but one example, Derrida examines the role that sound plays in Aristotle's definition of the difference between humans and animals: both can make sounds, Aristotle says, but only man, through his voice, is capable of meaning and reference (236–237).

7. In a significant study of passivity, albeit one that makes little mention of Derrida, Thomas Wall, referring to Levinas, links passivity to a "radical identification of the self with the Other that evacuates the self of sameness, stability, and self-certainty" (4).

Chapter 6

1. In *Animal Rites*, Wolfe contends that Jacques Derrida's sense of ethics "is diametrically opposed to what we find in a utilitarian like Peter Singer," whose "application of a 'calculable process,' namely, the utilitarian calculus" would effectively "tally up the 'interests' of the particular beings in question in a given situation, regardless of their species, and would determine what counts as a just act according to which action maximizes the greatest good for the greatest number." For Derrida, Wolfe writes, such an approach "reduces ethics to the very antithesis of ethics by reducing the aporia of judgment in which the possibility of justice resides to the mechanical unfolding of a positivist calculation" (69). This point, made by Derrida himself, is also picked up on by Kelly Oliver, as we discuss in the following chapter.

2. Twine clarifies in his book that "transhumanism" is a kind of "hyper-humanism," and "a troubling body of thought for posthumanism since it embraces and extends precisely the aspects of humanism that posthumanism has critiqued" (13). Transhumaism "extends values of liberal individualism, choice and bodily control," and it "recentres a valorization of the human" (13).

3. In *The Postmodern Explained*, Jean-François Lyotard defines "postmodernism" not as an epoch following the modern, but rather as "that which in the modern invokes the unrepresentable itself" (15), as that which, in other words, allows for differences that no representation can encompass or contain. For Derrida's objections to the term "postmodern," see his "Some Statements and Truisms."

4. The "presubjective" starting point is broached by Heidegger in his

Marburg Lecture course, as Derrida explicates in his reading of the
course in "*Geschlecht* I: sexual difference, ontological difference,"
where he notes that, for Heidegger, binary or dualistic difference is
but a secondary development, something introduced "later on," we
might say, by Western metaphysics across its theories of the subject or
self.

5. For a discussion of another feral child case, that of a boy found in the
forest of southern France near the village of Aveyron, a child attuned
only to animal sounds, thus without speech and beyond "raising" to
the civilized state, see McCance, "The Wild Child."

Chapter 7

1. Donovan and Adams identify Martha Nussbaum as representative of
Aristotelian feminism. Her *Frontiers of Justice: Disability, Nationality,
and Species Membership* presents a theory based on the idea of "capabilities" as what count in the ethical and political spheres.
2. The feminist care tradition, in line with Derrida's critique of "the
animal" abstraction, "recognizes the diversity of animals—one size
doesn't fit all" (3).

Chapter 8

1. In this interview, Singer cites the Zen roshi Philip Kapleau in favor
of Buddhist teachings of compassion and vegetarianism, and he
expresses the view that "Buddhists, in particular, should therefore
be playing an important role in the animal movement," and that he
would like to see "similar arguments [for compassion an vegetarian-
ism] developed within, say, the Hindu tradition" (617).
2. Edward Conze in *Buddhist Scriptures* cites one of Buddhaghosa's
explications in his *Visuddhimagga* of the implications of the First Precept even for a layperson. The passage, quoted here, suggests the tension in Buddhism between, on the one hand, refraining from doing
injury to living beings and, on the other hand, eating animals. Note
that the boy in this story recognizes the importance of noninjury,
but does not question whether a rabbit might be appropriate food.
"'Even those who have not formally undertaken to observe the

precepts may have the conviction that it is not right to offend against them. So it was with Cakkana, a Ceylonese boy. His mother was ill, and the doctor prescribed fresh rabbit meat for her. His brother sent him into the field to catch a rabbit, and he went as he was bidden. Now a rabbit had run into a field to eat of the corn, but in its eagerness to get there had got entangled in a snare, and gave forth cries of distress. Cakkana followed the sound, and thought: 'This rabbit has got caught there, and it will make a fine medicine for my mother!' But then he thought again: 'It is not suitable for me that, in order to preserve my mother's life, I should deprive someone else of his life.' And so he released the rabbit, and said to it: 'Run off, play with the other rabbits in the wood, eat grass and drink water!' On his return he told the story to his brother, who scolded him. He then went to his mother, and said to her: 'Even without having been told, I know quite clearly that I should not deliberately deprive any living being of life.' He then fervently resolved that these truthful words of his might make his mother well again, and so it actually happened'" (Conze 72–73).

3. See also Linzey's argument to this effect in his 2001 essay, "Vegetarianism as a Biblical Ideal."

4. On a similar note, Linzey, in answer to the question how a vegetarian Christian or Jew can rationalize Old Testament animal sacrifices, remarks that: "I'm not, of course, justifying the practice of animal sacrifice; but it's just conceivable that those who practiced animal sacrifice did not understand it a simply the gratuitous destruction of God's creatures; it was in some ways thought of as the liberation and the returning to God of that life back to the very life-source that caused everything to be." He adds that, from the Christian point of view, "the important thing theologically is that Jesus did not sacrifice animals" ("Conversation" 293).

Chapter 9

1. "If the post-Darwinian story of human origins that links human and animal being creates anxiety and ambivalence toward animality in modernist literature, it also opens up a space for critiques that trouble Western humanism and its abjection and repression of the animal" (Rohman, *Stalking the Subject*, 100).

Chapter 10

1. See also Nancy's *Noli me tangere: On the Raising of the Body* and his *Dis-enclosure: The Deconstruction of Christianity.* At stake in the latter book, rather, and one mark of its importance for critical animal studies, is the key question of "what the simple word *human* means," and with that goes the meaning of "humanism" (2). Behind the word *human*[*ism*], Nancy suggests, "behind what it says, behind what it hides—what it does not want to say, what it cannot or does not know how to say—stand the most imperious demands of thought today" (2).

2. See Leonard Lawlor's *This Is Not Sufficient: An Essay on Animality and Human Nature in Derrida* for an instance of such radical rethinking of (human and animal) life.

3. One thing distinguishing Ralph R. Acampora's *Corporal Compassion: Animal Ethics and Philosophy of Body* from a number of other approaches we discuss in previous chapters is his contention that the prevalent "bias in favor of mentality's moral significance" does not "get enough 'traction'" when it comes to including nonhuman species as worthy of moral consideration; it is not sufficient for bridging the gulf between humans and other creatures (4). As he puts it: "My approach suggests that this way of framing the issue has the experienced phenomena and the ethical problem entirely backwards" (4). For in his view, we do not find ourselves today "in some abstract, retro-Cartesian position of species solipsism where our minds seem to just float in a rarefied space of pure spectatorship apart from all ecological enmeshment and social connection with other organisms and persons, wondering, as it were, if 'there's anybody out there'" (4). Rather, we begin "always already caught up in the experience of being a live body thoroughly involved in a plethora of ecological and social interrelationships with other living bodies and people" (5). The ethical upshot of what he calls this "gestalt-shift" is, he suggests, "profound," in that, what now requires consideration is not "who gets included," or "who counts," but rather, just the other way round: what justifies dissociation and nonaffiliation (5)? Acampora approaches ethics "through the explication of cross-species relationships," joining the company of those thinkers "who see compassion rather than rationality at the root of interspecies morality" (23). He attempts to "show that cross-species compassion is mediated by

somatic experiences," experiences he labels *symphysis*, attempting with this word to emphasize the corporal component, and to distinguish "corporal compassion" from sympathy (23). The book is given primarily to the theorizing of this interspecies ethics of compassion, and to its application in the spheres of zoological parks and research laboratories.

Further to "pure spectatorship" and "having it backwards," Michael Naas, in "Derrida's Flair (For the Animals to Follow)," offers this remark: "In *The Animal That Therefore I Am* it is the animal that is first seen seeing and the human that is first seen—a simple reversal that is enough, claims Derrida, to reorient an entire philosophical tradition of thinking the animal other" (225).

Works Cited

Abbas, Fakhar. *Animal Rights in Islam: Islam and Animal's Rights*. Saarbrücken, Germany:VDM Verlag, 2009.

Acampora, Ralph R. *Corporal Compassion:Animal Ethics and Philosophy of Body*. Pittsburgh, PA: U of Pittsburgh P, 2006.

Adams, Carol J. *The Sexual Politics of Meat*. New York: Continuum, 1990.

———. *Neither Man nor Beast: Feminism and the Defense of Animals*. New York: Continuum, 1995.

———. *The Pornography of Meat*. New York: Continuum, 2004.

———. "'A Very Rare and Difficult Thing': Ecofeminism, Attention to Animal Suffering, and the Disappearance of the Subject." *A Communion of Subjects:Animals in Religion, Science, and Ethics*. Eds. Paul Waldau and Kimberley Patton. New York: Columbia UP, 2006. 591–604.

———. "The Rape of Animals, the Butchering of Women." *The Animal Ethics Reader*. Eds. Susan J. Armstrong and Richard G. Botzler. London and New York: Routledge, 2008. 268–273.

Adams, Carol J., and Josephine Donovan, eds. *Animals & Women: Feminist Theoretical Explorations*. Durham and London: Duke UP, 1995.

Agamben, Giorgio. *The Open: Man and Animal*. Trans. Kevin Attell. Stanford, CA: Stanford UP, 2004.

Akhtar, Aysha. "NASA's Wrong Stuff." *Huffington Post*. March 10, 2010.

Anderson, Elizabeth. "Animal Rights and the Value of Nonhuman Life." In *Animal Rights: Current Debates and New Directions*. Eds. Cass R. Sunstein and Martha C. Nussbaum. Oxford and New York: Oxford UP, 2004. 276–298.

Aristotle. *The Complete Works of Aristotle*. Vols. I and II. Ed. Jonathan Barnes. Princeton, NJ: Princeton UP, 1984.

Armstrong, Susan J., and Richard G. Botzler, eds. *The Animal Ethics Reader*. 2nd edition. New York and London: Routledge, 2008.

Bahn, Paul. *Journey through the Ice Age*. London: Weidenfeld & Nicolson, 1996.

Bailly, Jean-Christophe. *The Animal Side*. Trans. Catherine Porter. New York, NY: Fordham UP, 2011.

Baker, Steve. *Picturing the Beast: Animals, Identity and Representation*. Manchester and New York: Manchester UP, 1993.

———. *The Postmodern Animal*. London: Reaktion Books, 2000.

Bal, Mieke. *Reading Rembrandt: Beyond the Word-Image Opposition*. Cambridge, New York, Port Chester, Melbourne, Sydney: Cambridge UP, 1991.

———. *Double Exposures: The Subject of Cultural Analysis*. New York and London: Routledge, 1996.

———. Louise Bourgeois' *Spider: The Architecture of Art-Writing*. Chicago and London: U of Chicago P, 2001.

Ball, Matt. "Living and Working in Defense of Animals." *In Defense of Animals: The Second Wave*. Ed. Peter Singer. Oxford: Blackwell, 2006. 181–186.

Beattie, Heather, and Barbara Huck. *Wild West: Nature Living on the Edge. Endangered Species of Western North America*. Winnipeg, Canada: Heartland, 2009.

Beardsworth, Richard. *Derrida and the Political*. New York and London: Routledge, 1996.

Beauchamp, Tom, and James Childress. *Principles of Biomedical Ethics*. New York: Oxford UP, 1979.

Behnke, Elizabeth A. "From Merleau-Ponty's Concept of Nature to an Interspecies Practice of Peace." *Animal Others: On Ethics, Ontology, and Animal Life*. Ed. H. Peter Steeves. Albany: State U of New York P, 1999. 93–116.

Bekoff, Marc, and Jessica Pierce. *Wild Justice: The Moral Lives of Animals*. Chicago and London: U of Chicago P, 2009.

Berger, John. *About Looking*. New York: Pantheon Books, 1980.

———. *Why Look At Animals?* London and New York: Penguin, 2009.

Bermond, Bob. "The Myth of Animal Suffering." *The Animal Ethics Reader*. Eds. Susan J. Armstrong and Richard G. Botzler. New York and London: Routledge, 2003. 80–85.

Berry, Rynn. *Food for the Gods: Vegetarianism and the World's Religions*. New York: Pythagorean, 1998.

Birke, Lynda. *Feminism, Animals, and Science: The Naming of the Shrew*. Buckingham and Philadelphia: Open UP, 1994.

Birke, Lynda, and Luciana Parisi. "Animals, Becoming." *Animal Others: On Ethics, Ontology, and Animal Life*. Ed. H. Peter Steeves. Albany: State U of New York P, 1999. 55–73.

Bjerklie, Steve. "Size Matters: The Meat Industry and the Corruption of Darwinian Economics." *The CAFO Reader: The Tragedy of Industrial Animal Factories*. Ed. Daniel Imhoff. Berkeley and Los Angeles: The U of California P and the Foundation for Deep Ecology, 2010. 139–146.

Black's Law Dictionary. 4th edition. St Paul, MN: West, 1968.

Borges, Jorge Luis. *The Book of Imaginary Beings*. Trans. Norman Thomas di Giovanni. New York: Dutton, 1970.

Buddhaghosa, Bhadantacariya. *The Path to Purification (Visuddhimagga)*. Trans. Bhikkhu Nanomoli. Onalaska, MA: Pariyatti, 2003.

Burt, Jonathan. "The Illumination of the Animal Kingdom: The Role of Light and Electricity in Animal Representation." *The Animals Reader: The Essential Classic and Contemporary Writings*. Eds. Linda Kaloff and Amy Fitzgerald. Oxford and New York: Berg, 2007. 289–301.

Calarco, Matthew. *Zoographies: The Question of the Animal from Heidegger to Derrida*. New York: Columbia UP, 2008.

———. "Toward An Agnostic Animal Ethics." Paola Cavalieri, *The Death of the Animal*. New York: Columbia UP, 2009. 75–84.

Capra, Fritjof. *The Tao of Physics: An Exploration of the Parallels between Modern Physics and Eastern Mysticism*. Boston: Shambhala, 1975.

———. *The Turning Point: Science, Society, and the Rising Culture*. New York: Bantam, 1982.

Carlino, Andrea. *Books of the Body: Anatomical Ritual and Renaissance Learning*. Trans. John Tedeschi and Anne C. Tedeschi. Chicago: U of Chicago P, 1999.

Castricano, Jodey, ed. *Animal Subjects: An Ethical Reader in a Posthuman World*. Waterloo: Wilfrid Laurier UP, 2008.

Cavalieri, Paola. *The Animal Question: Why Nonhuman Animals Deserve Human Rights*. Oxford and New York: Oxford UP, 2001.

———. "A Missed Opportunity: Humanism, Anti-humanism and the Animal Question." In *Animal Subjects: An Ethical Reader in a Posthuman World*. Ed. Jodey Castricano. Waterloo: Wilfrid Laurier UP, 2008. 97–123.

————. *The Death of the Animal: A Dialogue.* New York: Columbia UP, 2009.

Cavalieri, Paola, and Peter Singer, eds. *The Great Ape Project: Equality Beyond Humanity.* New York: St. Martin's Griffin, 1993.

Cavell, Stanley, Cora Diamond, John McDowell, Ian Hacking, and Cary Wolfe. *Philosophy & Animal Life.* New York: Columbia UP, 2008.

Chapple, Christopher Key. *Nonviolence to Animals, Earth, and Self in Asian Traditions.* Albany: State U of New York P, 1993.

————. "Inherent Value without Nostalgia: Animals and the Jaina Tradition." *A Communion of Subjects: Animals in Religion, Science, and Ethics.* Eds. Paul Waldau and Kimberley Patton. New York: Columbia UP, 2006. 241–249.

Cixous, Hélène, and Catherine Clément. "Sorties: Out and Out: Attacks/Ways Out/Forays." *The Newly Born Woman.* Trans. Betsy Wing. Minneapolis: U of Minnesota P, 1986. 63–132.

Clark, David L. "On Being 'the last Kantian in Nazi Germany': Dwelling with Animals after Levinas." *Animal Acts: Configuring the Human in Western History.* Eds. Jennifer Ham and Matthew Senior. New York and London: Routledge, 1997. 165–198.

Coetzee, J. M. *The Lives of Animals.* Princeton, NJ: Princeton University Press, 1999.

————. *Elizabeth Costello.* London: Vintage, 1999.

Conze, Edward, ed. and trans. *Buddhist Scriptures.* London: Penguin, 1959.

Cook, Christopher D. "Sliced and Diced: The Labor You Eat." *The CAFO Reader: The Tragedy of Industrial Animal Factories.* Ed. Daniel Imhoff. Berkeley and Los Angeles: U of California P and the Foundation for Deep Ecology, 2010. 232–239.

Cross, John A. "Change in America's Dairyland." *Geographical Review* 91, 4 (October 2001): 702–714.

Darwin, Charles. *On the Origin of Species By Means of Natural Selection.* Cambridge, MA: Oxford UP, 2009.

————. *The Descent of Man.* London and New York: Penguin Books, 2004.

Datson, Lorraine, and Gregg Mitman, eds. *Thinking with Animals: New Perspectives on Anthropomorphism.* New York: Columbia UP, 2006.

Dawkins, Marian Stamp. "What Is Good Welfare and How Can We Achieve It?" *The Future of Animal Farming: Renewing the Ancient Contract.* Eds. Marian Stamp Dawkins and Roland Bonney. Malden, MA,

Oxford, and Victoria, Australia: Blackwell, 2008. 73–82.

Dawkins, Marian Stamp, and Roland Bonney, eds. *The Future of Animal Farming: Renewing the Ancient Contract*. Malden, MA, Oxford, and Victoria, Australia: Blackwell, 2008.

de Bary, Wm. Theodore, ed. *Sources of Indian Tradition*. New York: Columbia UP, 1958.

DeGrazia, David. "Animals for Food." *The Animal Ethics Reader*. Eds. Susan J. Armstrong and Richard G. Botzler. London and New York: Routledge, 2008. 219–224.

Dekoven, Marianne. "Why Animals Now?" *PMLA* 124, 2 (March 2009): 361–369.

Derrida, Jacques. "Differance." *Speech and Phenomena: And Other Essays on Husserl's Theory of Signs*. Trans. David B. Allison. Evanston, IL: Northwestern UP, 1973. 129–160.

———. *Of Grammatology*. Trans. Gayatri Spivak. Baltimore, MD: Johns Hopkins UP, 1974.

———. "Plato's Pharmacy." Trans. Barbara Johnson. *Dissemination*. Chicago, IL: U of Chicago P, 1981. 61–171.

———. "White Mythology: Metaphor in the Text of Philosophy." Trans. Alan Bass. *Margins of Philosophy*. Chicago, IL: U of Chicago P, 1982. 207–271.

———. "Geschlecht: sexual difference, ontological difference." Trans. Ruben Berezdivin. *Research in Phenomenology* 13 (1983): 65–83.

———. *Glas*. Trans. John P. Leavey Jr., and Richard Rand. Lincoln and London: U of Nebraska P, 1986.

———. "Geschlecht II: Heidegger's Hand." Trans. John P. Leavey Jr. *Deconstruction and Philosophy: The Texts of Jacques Derrida*. Ed. John Sallis. Chicago: U of Chicago P, 1987. 161–196.

———. *Of Spirit: Heidegger and the Question*. Trans. Geoffrey Bennington and Rachel Bowlby. Chicago, IL: U of Chicago P, 1989.

———. "Some Statements and Truisms About Neologisms, Newisms, Postisms, Parasitisms, and Other Small Seismisms." Trans. Anne Tomiche. *The States of Theory: History, Art, and Critical Discourse*. Ed. David Carroll. New York: Columbia UP, 1990. 63–94.

———. "'Eating Well,' or the Calculation of the Subject: An Interview with Jacques Derrida." Trans. Peter Connor and Avital Ronnell. *Who Comes After the Subject?* Eds. Eduardo Cadava, Peter Connor, Jean-Luc Nancy. London: Routledge, 1991. 96–119.

————. "Passions: 'An Oblique Offering.'" Trans. David Wood. *Derrida: A Critical Reader*. Ed. David Woods. Cambridge, MA: Blackwell, 1992. 5–35.

————. "Mochlos; or, The Conflict of the Faculties." Trans. Richard Rand and Amy Wygant. *Logomachia: The Conflict of the Faculties*. Ed. Richard Rand. Lincoln: U of Nebraska P, 1992. 3–34.

————. "Heidegger's Ear: Philopolemology (Geschlecht IV)." Trans. John P. Leavey Jr. *Reading Heidegger: Commemorations*. Ed. John Sallis. Bloomington: Indiana UP, 1993. 163–218.

————. "A Silkworm of One's Own." Trans. Geoffrey Bennington. *Oxford Literary Review* 18 (1997): 3–63.

————. *On Touching: Jean-Luc Nancy*. Trans. Christine Irizarry. Stanford, CA: Stanford UP, 2000.

————. "Faith and Knowledge: Two Sources of 'Religion' at the Limits of Reason Alone." Trans. Samuel Weber. *Acts of Religion*. Ed. Gil Anidjar. London: Routledge, 2002. 40–101.

————. "Autoimmunity: Real and Symbolic Suicides: A Dialogue with Jacques Derrida." *Philosophy in a Time of Terror: Dialogues with Jürgen Habermas and Jacques Derrida. Giovanna Boradorri*. Chicago and London: U of Chicago P, 2003. 85–136.

————. *Rogues: Two Essays on Reason*. Trans. Pascale-Anne Brault and Michael Naas. Stanford, CA: Stanford UP, 2005.

————. *The Animal That Therefore I Am*. Trans. David Wills. New York: Fordham UP, 2008.

————. *The Beast and the Sovereign, Volume 1*. Trans. Geoffrey Bennington. Chicago and London: U of Chicago P, 2009.

————. *The Beast and the Sovereign, Volume II*. Trans. Geoffrey Bennington. Chicago and London: U of Chicago P, 2011.

Derrida, Jacques, and Elisabeth Roudinesco. "Violence Against Animals." Trans. Jeff Fort. *For What Tomorrow?* Stanford, CA: Stanford UP, 2004. 62–76.

Derrida, Jacques, and Peter Eisenman. *Chora L Works*. Eds. Jeffrey Kipnis and Thomas Leeser. New York: Monacelli Press, 1997.

Descartes, René. *Discourse on the Method*. In *The Philosophical Writings of Descartes*. Ed. and trans. John Cottingham, Robert Stoothoff, Dugald Murdoch. Vol 1. Cambridge: Cambridge UP, 1985. 111–151.

————. *Description of the Human Body and of All Its Functions. The Philosophical Writings of Descartes*. Ed. and trans. John Cottingham, Robert Stoothoff, Dugald Murdoch. Vol 1. Cambridge: Cambridge UP, 1985. 314–324.

————. "Letter to the Marquess of Newcastle, 23 November 1646." *The Philosophical Writings of Descartes: The Correspondence.* Ed. and trans. John Cottingham, Robert Stoothoff, Dugald Murdoch, and Anthony Kenny. Cambridge: Cambridge UP, 1991. 302–304.

————. "Letter to More, 5 February 1649." *The Philosophical Writings of Descartes: The Correspondence.* Ed. and trans. John Cottingham, Robert Stoothoff, Dugald Murdoch, and Anthony Kenny. Cambridge: Cambridge UP, 1991. 360–67.

D'Silva, Joyce. "The Urgency of Change: A View from a Campaigning Organization." *The Future of Animal Farming: Renewing the Ancient Contract.* Eds. Marian Stamp Dawkins and Roland Bonney. Malden, MA, Oxford, and Victoria, Australia: Blackwell, 2008. 33–44.

Doniger, Wendy. "Hinduism By Any Other Name." *Wilson Quarterly* 15, 3 (Summer 1991): 35–41.

————. "Reflections." *The Lives of Animals.* J. M. Coetzee. Princeton, NJ: Princeton UP, 1999. 93–106.

Donovan, Josephine, and Carol J. Adams, eds. *Beyond Animal Rights: A Feminist Caring Ethic for the Treatment of Animals.* New York: Continuum, 1996.

————. *The Feminist Care Tradition in Animal Ethics.* New York: Columbia UP, 2007.

Ehrenfeld, David. "Foreword." *Ethics on the Ark: Zoos, Animal Welfare, and Wildlife Conservation.* Eds. Bryan G. Norton, Michael Hutchins, Elizabeth F. Stevens, and Terry L. Maple. Washington and London: Smithsonian Institution Press, 1995. xvii–xix.

Eisnitz, Gail A. *Slaughterhouse.* New York: Prometheus Books, 2007.

Ellul, Jacques. *The Technological Society.* New York: Knopf, 1964.

————. *The Technological System.* New York: Seabury Press, 1980.

Foltz, Richard C. *Animals in Islamic Tradition and Muslim Cultures.* Oxford: Oneworld, 2006.

Forward, Martin, and Mohamed Alam. "Islam." *The Animal Ethics Reader.* Eds. Susan J. Armstrong and Richard G. Botzler. London and New York: Routledge, 2008. 294–296.

Foucault, Michel. *The Order of Things: An Archaeology of the Human Sciences.* Trans. unidentified collective. New York: Random House, 1973.

————. *The History of Sexuality, Volume 1: An Introduction.* New York: Vintage, 1980.

Francione, Gary L. *Animals, Property, and the Law.* Philadelphia: Temple UP, 1995.

Franklin, Sarah. "Dolly's Body: Gender, Genetics, and the New Genetic

Capital." *The Animals Reader: The Essential Classic and Contemporary Writings*. Eds. Linda Kalof and Amy Fitzgerald. Oxford and New York: Berg, 2007. 349–361.

Garber, Majorie. "Reflections." *The Lives of Animals*. J. M. Coetzee. Princeton, NJ: Princeton UP, 1999. 71–84.

Garner, Robert. *Animal Ethics*. Cambridge, UK: Polity Press, 2005.

Gilligan, Carol. *In a Different Voice: Psychological Theory and Women's Development*. Cambridge, MA: Harvard UP, 1982.

Goodall, Jane. *In the Shadow of Man*. New York: Mariner Books, 2000.

———. *Through a Window*. New York: Mariner Books, 2000.

———. *My Life with the Chimpanzees*. New York: Aladdin, 2002.

Goodall, Jane, with Thane Maynard and Gail Hudson. *Hope for Animals and Their World: How Endangered Species Are Being Rescued from the Brink*. New York and Boston: Grand Central, 2009.

Grant, George. *Technology and Empire: Perspectives on North America*. Toronto: Anansi, 1991.

Green, Jonathon. *Cassell's Dictionary of Slang*. London: Cassell, 1998.

Gregor, Howard F. "Industrialized Drylot Dairying: An Overview." *Economic Geography* 39, 4 (October 1963): 299–318.

Grenander, M. E. "Review: The Sexual Politics of Meat." *NWSA Journal* 3, 2 (Spring 1991): 335–336.

Griffiths, Huw, Ingrid Poulter, and David Sibley. "Feral Cats in the City." *Animal Spaces, Beastly Places: New Geographies of Human-Animal Relations*. Eds. Chris Philo and Chris Wilbert. New York: Routledge, 2000. 56–70.

Gross, Charles. "Galen and the Squealing Pig." *Neuroscientist* 4 (1998): 216–221.

Gruen, Lori. *Ethics and Animals: An Introduction*. Cambridge, UK, New York, Melbourne: Cambridge UP, 2011.

Guerrini, Anita. *Experimenting with Humans and Animals: From Galen to Animal Rights*. Baltimore and London: Johns Hopkins UP, 2003.

Gullo, Andrea, Unna Lassiter, and Jennifer Wolch. "The Cougar's Tale." *Animal Geographies: Place, Politics, and Identity in the Nature-Culture Borderlands*. Eds. Jennifer Wolch and Jody Emel. London: Verso, 1998. 139–161.

Haraway, Donna, J. *Primate Visions: Gender, Race, and Nature in the World of Modern Science*. New York and London: Routledge, 1989.

———. *Simians, Cyborgs and Women: The Reinvention of Nature*. New York and London: Routledge, 1991.

———. *The Companion Species Manifesto: Dogs, People, and Significant Otherness*. Chicago: Prickly Paradigm Press, 2003.

———. *When Species Meet*. Minneapolis and London: U of Minnesota P, 2008.

Harding, Sandra. *The Science Question in Feminism*. Ithaca: Cornell UP, 1986.

Harris, Ian. "'A vast unsupervised recycling plant': Animals and the Buddhist Cosmos." *A Communion of Subjects: Animals in Religion, Science, and Ethics*. Eds. Paul Waldau and Kimberley Patton. New York: Columbia UP, 2006. 207–217.

Hegel, G. W. F. *Lectures on the Philosophy of Religion: One Volume Edition*. Trans. R. F. Brown, P. C. Hodgson, and J. M. Stewart. Ed. Peter C. Hodgson. Berkeley, Los Angeles, London: U of California P, 1988.

Heidegger, Martin. *The Fundamental Concepts of Metaphysics: World, Finitude, Solitude*. Trans. William McNeil and Nicholas Walker. Bloomington and Indianapolis: Indiana UP, 1995.

Heisenberg, Werner. *Physics and Beyond*. New York: Harper & Row, 1972.

Holloway, Lewis, Carol Morris, Ben Gilna, and David Gibbs. "Biopower, Genetics and Livestock Breeding: (Re)Constituting Animal Populations and Heterogeneous Biosocial Collectivities." *Transactions of the Institute of British Geographers*, New Series 34, 3 (July 2009): 394–407.

Huggan, Graham, and Helen Tiffin. *Postcolonial Criticism: Literature, Animals, Environment*. New York and London: Routledge, 2010.

Husserl, Edmund. *Cartesian Meditations: An Introduction to Phenomenology*. Trans. Dorion Cairns. The Hague: Martinus Nijhoff, 1960.

Imhoff, Daniel, ed. *The CAFO Reader: The Tragedy of Industrial Animal Factories*. Berkeley and Los Angeles: The U of California P and the Foundation for Deep Ecology, 2010.

Ingold, T. "From Trust to Domination: An Alternate History of Human-Animal Relations." In *Animals and Human Society*. Eds. A. Manning and J. Serpell. London and New York: Routledge, 1994.

Ingraham, Catherine. *Architecture and the Burdens of Linearity*. New Haven and London: Yale UP, 1998.

———. *Architecture Animal Human: The Asymmetrical Condition*. London and New York: Routledge, 2006.

Issacs, Ronald H. *Animals in Jewish Thought and Tradition*. Lanham, MD: Jason Aronson, 2000.

Jamieson, Dale. "Against Zoos." *In Defense of Animals: The Second Wave*. Ed. Peter Singer. Oxford: Blackwell, 2006. 132–143.

Jay, Martin. "Scopic Regimes of Modernity." *Vision and Visuality*. Ed. Hal Foster. Seattle: Bay Press, 1988. 4–23.

———. *Downcast Eyes: The Denigration of Vision in Twentieth-Century Thought*. Berkeley: U of California P, 1993.

Kafka, Franz. *The Metamorphosis*. Trans. Stanley Comgold. Toronto, New York, London, Sydney: Bantam Books, 1972.

Kalechofsky, Roberta. "Hierarchy, Kinship, and Responsibility: The Jewish Relationship to the Animal World." *A Communion of Subjects: Animals in Religion, Science, and Ethics*. Eds. Paul Waldau and Kimberley Patton. New York: Columbia UP, 2006. 91–99.

Kant, Immanuel. *The Conflict of the Faculties*. Trans. Mary J. Gregor. Lincoln: U of Nebraska P, 1979.

———. *Anthropology from a Pragmatic Point of View*. Trans. Robert B. Louden. Cambridge, UK: Cambridge UP, 2006.

Keller, Evelyn Fox. *Secrets of Life, Secrets of Death: Essays on Language, Gender, and Science*. London and New York: Routledge, 1992.

Koestler, Arthur. *The Ghost in the Machine*. London: Picador, 1967.

Kristeva, Julia. *Powers of Horror: An Essay on Abjection*. Trans. Leon Roudiez. New York: Columbia UP, 1982.

———. *Revolution in Poetic Language*. Trans. Margaret Waller. New York: Columbia UP, 1984.

———. *Tales of Love*. Trans. Leon Roudiez. New York: Columbia UP, 1987.

Kuzniar, Alice A. *Melancholia's Dog: Reflections on Our Animal Kinship*. Chicago and London: U of Chicago P, 2006.

Lacan, Jacques. "The Function and Field of Speech and Language in Psychoanalysis." Trans. Alan Sheridan. *Écrits: A Selection*. New York: Norton, 1977. 30–113.

Lawlor, Leonard. *This Is Not Sufficient: An Essay on Animality and Human Nature in Derrida*. New York: Columbia UP, 2007.

Leopold, Aldo. "The State of the Profession." *Journal of Wildlife Management* 4 (1940): 343–346.

———. *A Sand County Almanac*. New York: Ballantine Books/Random House, (1949), 1986.

Leroi-Gourhan, André. *Gesture and Speech*. Trans. Anna Bostock Berger. Cambridge, MA and London, England: MIT Press, 1993.

Levinas, Emmanuel. "The Name of a Dog, or Natural Rights." Trans. Seán Hand. *Difficult Freedom: Essays on Judaism*. Baltimore: Johns Hopkins University Press, 1990. 151–153.

Lingis, Alphonso. "Bestiality." *Animal Others: On Ethics, Ontology, and Animal Life*. Ed. H. Peter Steeves. Albany: State U of New York P, 1999. 37–54.

———. "Animal Body, Inhuman Face." *Zoontologies: The Question of the Animal*. Ed. Cary Wolfe. Minneapolis: U of Minnesota P, 2003. 165–182.

———. "Inner Space." In *Mosaic: A Journal for the Interdisciplinary Study of Literature* 43, 2 (June 2010): 37–44.

Linzey, Andrew. *Christianity and the Rights of Animals*. London: SPCK, 1987.

———. "Conversation: Protestant Christianity." *Rynn Berry, Food for the Gods: Vegetarianism and the World's Religions*. New York: Pythagorean Press, 1998. 271–301.

———. "Vegetarianism as a Biblical Ideal." *Religious Vegetarianism: From Hesiod to the Dalai Lama*. Eds. Kerry S. Walters and Lisa Portmess, Albany: State U of New York P, 2001. 126–139.

———. "The Bible and Killing for Food." In *The Animal Ethics Reader*. Eds. Susan J. Armstrong and Richard G. Botzler. London and New York: Routledge, 2008. 286–293.

———. *Why Animal Suffering Matters: Philosophy, Theology, and Practical Ethics*. Oxford and New York: Oxford UP, 2009.

Linzey, Andrew, ed. *The Link Between Animal Abuse and Human Violence*. Brighton and Portland: Sussex Academic Press, 2009.

Lobao, Linda, and Katherine Meyer. "The Great Agricultural Transition: Crisis, Change, and Social Consequences of Twentieth-Century US Farming." *Annual Review of Sociology* 27 (2001): 103–124.

Lynn, William S. "Animals, Ethics, and Geography." *Animal Geographies: Place, Politics, and Identity in the Nature-Culture Borderlands*. Eds. Jennifer Wolch and Jody Emel. London: Verso, 1998. 280–297.

Lyotard, Jean-François. *The Postmodern Explained*. Trans. Don Barry et al. Minneapolis and London: U of Minnesota P, 1992.

Malamud, Randy. *Reading Zoos: Representations of Animals in Captivity*. New York: New York University Press, 2008.

Mallin, Michael A., and Lawrence B. Cahoon. "Industrialized Animal Production: A Major Source of Nutrient and Microbial Pollution to Aquatic Ecosystems." *Population and Environment* 24, 5 (May 2003): 369–385.

Marcus, Erik. *Meat Market: Animals, Ethics, and Money*. Boston: Brio Press, 2005.

Martel, Yann. *Life of Pi.* Toronto: Vintage Canada, 2001.

Martel, Yann. *Beatrice & Virgil.* Toronto: Random House Canada, 2010.

Mason, Jim. *An Unnatural Order: The Roots of our Destruction of Nature.* New York: Lantern Books, 2005.

Mason, Jim, and Peter Singer. *Animal Factories.* New York: Crown, 1980.

Masri, Al-Hafiz Basheer Ahmad. *Animal Welfare in Islam.* Markfield, Leicestershire, UK: Islamic Foundation, 2009.

Matheny, Gaverick. "Utilitarianism and Animals." *In Defense of Animals: The Second Wave.* Ed. Peter Singer. Oxford: Blackwell, 2006. 13–25.

McCance, Dawne. *Medusa's Ear: University Foundings from Kant to Chora L.* Albany: State U of New York P, 2004.

———. "Anatomy as Speech Act: Vesalius, Descartes, Rembrandt or, The Question of 'the animal' in the Early Modern Anatomy Lesson." *Animal Subjects: An Ethical Reader in a Posthuman World.* Ed. Jodey Castricano. Waterloo: Wilfrid Laurier UP, 2008. 63–95.

———. "The Wild Child." *Canadian Journal of Film Studies: Revue Canadienne D'Études Cinématographiques* 17, 1 (Spring 2008): 69–80.

Meine, Curt D. "Moving Mountains: Aldo Leopold and a Sand County Alamanac." *Aldo Leopold and the Ecological Conscience.* Eds. Richard L. Knight and Suzanne Riedel. Oxford and New York: Oxford UP, 2002. 14–31.

Midgley, Mary. *Animals and Why They Matter.* Athens: U of Georgia P, 1983.

———. *Science as Salvation: A Modern Myth and Its Meaning.* London and New York: Routledge, 1994.

———. *Evolution as a Religion.* London and New York: Routledge, 2002.

———. *Beast and Man: The Roots of Human Nature.* London and New York: Routledge, 2002.

———. *The Myths We Live By.* London and New York: Routledge, 2004.

———. "Why Farm Animals Matter." *The Future of Animal Farming: Renewing the Ancient Contract.* Eds. Marian Stamp Dawkins and Roland Bonney. Malden, MA, Oxford, and Victoria, Australia: Blackwell, 2008. 21–31.

Miles, H. Lyn White. "Language and the Orang-utan: The Old 'Person' of the Forest." *The Great Ape Project: Equality Beyond Humanity.* Eds. Paola Cavalieri and Peter Singer. New York: St. Martin's Griffin, 1993. 42–57.

Miller, Debra, ed. *Factory Farming.* Farmington Hills, MI: Greenhaven Press, 2010.

Miller, Harlan B. "The Wahokies." *The Great Ape Project: Equality Beyond Humanity*. Eds. Paola Cavalieri and Peter Singer. New York: St. Martin's Griffin, 1993. 230–236.

Millett, Kate. *Sexual Politics*. London: Virago, 1969, 1977.

Mitchell, Robert W. "Humans, Nonhumans and Personhood." *The Great Ape Project: Equality Beyond Humanity*. Eds. Paola Cavalieri and Peter Singer. New York: St. Martin's Griffin, 1993. 237–247.

Mitchell, W. J. T. "Foreword: The Rights of Things." *Cary Wolfe, Animal Rites: American Culture, the Discourse of Species, and Posthumanist Theory*. Chicago and London: U of Chicago P, 2003. ix–xiv.

Mithen, Steven. "The Evolution of Imagination: An Archaeological Perspective." *SubStance* Vol. 30, 1–2 (2001): 28–54.

———. "The Hunter-Gatherer Prehistory of Human-Animal Relations." *The Animals Reader: The Essential Classic and Contemporary Writings*. Eds. Linda Kaloff and Amy Fitzgerald. Oxford and New York: Berg, 2007. 117–128.

Mittelstaedt, Martin. "Eureka! Less Really Is More—Deadly." *The Globe and Mail*. March 12, 2010, F1–7.

Moi, Toril. *Sexual/Textual Politics: Feminist Literary Theory*. London and New York: Methuen, 1985.

Naas, Michael. "Derrida's Flair (For the Animals to Follow . . .)." *Research In Phenomenology* 40, 2 (2010): 219–242.

Nancy, Jean-Luc. "Unum Quid." *Ego Sum. Paris: Flammarion, 1979*. 129–164.

———. *Dis-Enclosure: The Deconstruction of Christianity*. Trans. Bettina Bergo, Gabriel Malenfant, and Michael B. Smith. New York: Fordham UP, 2008.

———. *Noli Me Tangere: On the Raising of the Body*. Trans. Sarah Clift, Pascale-Anne Brault and Michael Naas. New York: Fordham UP, 2008.

———. *Corpus*. Trans. Richard A. Rand. New York: Fordham UP, 2008.

Nierenberg, Danielle. *Happier Meals: Rethinking the Global Meat Industry*. Washington DC: WorldWatch, 2005.

Norton, Bryan G., Michael Hutchins, Elizabeth F. Stevens, and Terry L. Maple, eds. *Ethics on The Ark: Zoos, Animal Welfare, and Wildlife Conservation*. Washington and London: Smithsonian Institution Press, 1995.

Nussbaum, Martha C. *Frontiers of Justice: Disability, Nationality, Species Membership*. Cambridge, MA and London, England: Belknap Press, 2006.

Oliver, Kelly. *Family Values: Subjects Between Nature and Culture.* New York and London: Routledge, 1997.

———. *Animal Lessons: How They Teach Us to Be Human.* New York: Columbia UP, 2009.

O'Malley, C. D. *Andreas Vesalius of Brussels 1514–1564.* Berkeley and Los Angeles: U of California P, 1964.

Oxford English Dictionary. Complete Text, Compact Edition. Two Volumes. Oxford and New York: Oxford UP, 1971; 1985.

Panofsky, Erwin. *Perspective as Symbolic Form.* Trans. Christopher S. Wood. New York: Zone Books, 1991.

Partridge, Eric. *Origins: A Short Etymological Dictionary of Modern English.* New York: Macmillan, 1958.

Patton, Kimberley. "Animal Sacrifice: Metaphysics of the Sublimated Victim." *A Communion of Subjects: Animals in Religion, Science, and Ethics.* Eds. Paul Waldau and Kimberley Patton. New York: Columbia UP, 2006. 391–405.

———. "'Caught with Ourselves in the Net of Life and Time': Traditional Views of Animals in Religion." *A Communion of Subjects: Animals in Religion, Science, and Ethics.* Eds. Paul Waldau and Kimberley Patton. New York: Columbia UP, 2006. 27–39.

Perlo, Katherine Wills. *Kinship and Killing: The Animal in World Religions.* New York: Columbia UP, 2009.

Peterson, Dale. *Eating Apes.* Berkeley, Los Angeles, London: U of California P, 2003.

———. *The Moral Lives of Animals.* New York, Berlin, London, Sydney: Bloomsbury Press, 2011.

Philo, Chris, and Chris Wilbert. "Animal spaces, beastly places: An introduction." *Animal Spaces, Beastly Places: New Geographies of Human-Animal Relations.* Eds. Chris Philo and Chris Wilbert. London and New York: Routledge, 2000. 1–34.

Philo, Chris. "Animals, Geography, and the City: Notes on Inclusion and Exclusion." *Animal Geographies: Place, Politics, and Identity in the Nature-Culture Borderlands.* Eds. Jennifer Wolch and Jody Emel. London: Verson, 1998. 51–71.

Pirsig, Robert. *Zen and the Art of Motorcycle Maintenance: An Inquiry into Values.* New York: William Morrow, 1974.

Rachels, James. "The Basic Argument for Vegetarianism." *The Animal Ethics Reader.* Eds. Susan J. Armstrong and Richard G. Botzler. London and New York: Routledge, 2008. 260–267.

Raloff, Janet. "Dying Breeds." *Science News* 152, 14 (October 4, 1997): 216–218.

Rand, Richard. "Preface." *Logomachia: The Conflict of the Faculties*. Ed. Richard Rand. Lincoln: U of Nebraska P, 1992. vii–xii.

Rawles, Kate. "Environmental Ethics and Animal Welfare: Re-Forging a Necessary Alliance." *The Future of Animal Farming: Renewing the Ancient Contract*. Eds. Marian Stamp Dawkins and Roland Bonney. Malden, MA, Oxford, and Victoria, Australia: Blackwell, 2008. 45–62.

Regan, Tom. *The Case for Animal Rights*. 3rd edition. Berkeley and Los Angeles: U of California P, 1983, 2004.

———. ed. *Animal Sacrifices: Religious Perspectives on the Use of Animals in Science*. Philadelphia: Temple UP, 1986.

———. "Are Zoos Morally Defensible?" *Ethics on the Ark: Zoos, Animal Welfare, and Wildlife Conservation*. Eds. Bryan G. Norton et al. Washington and London: Smithsonian Institution Press, 1995. 38–51.

———. "Foreword." *Animal Others: On Ethics, Ontology, and Animal Life*. Ed. H. Peter Steeves. Albany: State U of New York P, 1999. xi–xiii.

———. *Defending Animal Rights*. Urbana and Chicago: U of Illinois P, 2001.

———. *Empty Cages: Facing the Challenge of Animal Rights*. Lanham, MD: Rowman & Littlefield, 2004.

Reiss, Timothy, J. *The Discourse of Modernism*. Ithaca and London: Cornell UP, 1982.

Rhodes, V. James. "The Industrialization of Hog Production." *Review of Agricultural Economics* 17 (1995): 107–118.

Robbins-Roth, Cynthia. *From Alchemy to IPO: The Business of Biotechnology*. New York: Basic Books, 2000.

Roberts, Mark S. *The Mark of the Beast: Animality and Human Oppression*. West Lafayette, IN: Purdue UP, 2008.

Rohman, Carrie. *Stalking the Subject: Modernism and the Animal*. New York: Columbia UP, 2009.

Rollin, Bernard. "The Ethics of Agriculture: The End of True Husbandry." *The Future of Animal Farming: Renewing the Ancient Contract*. Eds. Marian Stamp Dawkins and Roland Bonney. Malden, MA, Oxford, and Victoria, Australia: Blackwell, 2008. 7–19.

Roof, Judith. "From Protista to DNA (and Back Again): Freud's Psychoanalysis of the Single-Celled Organism." *Zootologies: The Question of the Animal*. Ed. Cary Wolfe. Minneapolis and London: U of Minnesota P, 2003. 101–120.

Rosen, Steve. *Diet for Transcendence: Vegetarianism and the World Religions.* New Delhi: New Age Books, 1997.

Rowlands, Mark. *Animals Like Us.* London: Verso, 2002.

Royle, Nicholas. *Veering: A Theory of Literature.* Edinburgh, Scotland: Edinburgh UP, 2011.

Said, Edward W. *Orientalism.* New York: Vintage Books, 1979.

Santmire, H. Paul. *The Travail of Nature: The Ambiguous Ecological Promise of Christian Theology.* Philadelphia: Fortress Press, 1985.

Schaffner, Joan E. *An Introduction to Animals and the Law.* London and New York: Palgrave Macmillan, 2011.

Schlosser, Eric. *Fast Food Nation: The Dark Side of the All-American Meal.* New York: Houghton Mifflin, 2001.

Scholtmeijer, Marian. "What is 'Human'? Metaphysics and Zoontology in Flaubert and Kafka." *Animal Acts: Configuring the Human in Western History.* Eds. Jennifer Ham and Matthew Senior. New York and London: Routledge, 1997. 127–143.

Shideler, James H. "The Farm in American History." *Magazine of History* 5, 3 (Winter 1991): 18–23.

Slifkin, Natan. *Man and Beast: Our Relationships with Animals in Jewish Law and Thought.* Brooklyn, NY: Yashar Books, 2006.

Singer, Peter. "Animal Experimentation: Philosophical Perspectives." *The Encyclopedia of Bioethics.* Ed. Warren T. Reich. New York: Free Press, 1978. 79–83.

———. *Practical Ethics.* Cambridge, New York, Melbourne: Cambridge UP, 1979.

———. "Reflection." J. M. Coetzee, *The Lives of Animals.* Princeton, NJ: Princeton UP, 1999. 85–91.

———. Ed. *In Defense of Animals: The Second Wave.* Oxford: Blackwell, 2006.

———. "Animal Protection and the Problem of Religion: An Interview with Peter Singer." *A Communion of Subjects: Animals in Religion, Science, and Ethics.* Eds. Paul Waldau and Kimberley Patton. New York: Columbia UP, 2006. 616–618.

———. "Foreword." *The Future of Animal Farming: Renewing the Ancient Contract.* Eds. Marian Stamp Dawkins and Roland Bonney. Malden, MA, Oxford, and Victoria, Australia: Blackwell, 2008. vii–ix.

———. *Animal Liberation.* 4th edition. New York and London: Harper-Collins, 2009.

————. "Foreword." Paola Cavalieri, *The Death of the Animal*. New York: Columbia UP, 2009. ix–xii.

Singer, Peter, and Jim Mason. *The Ethics of What We Eat: Why Our Food Choices Matter*. Kutztown, PA: Rodale, 2006.

Slicer, Deborah. "Your Daughter or Your Dog? A Feminist Reassessment of the Animal Research Issue." *The Feminist Care Tradition in Animal Ethics*. Ed. Josephine Donovan and Carol J. Adams. New York: Columbia UP, 2007. 105–124.

Smuts, Barbara. "Reflections." *The Lives of Animals*. J. M. Coetzee. Princeton, NJ: Princeton UP, 1999. 107–120.

Steeves, H. Peter, ed. *Animal Others: On Ethics, Ontology, and Animal Life*. Albany: State U of New York P, 1999.

————. "Introduction." *Animal Others: On Ethics, Ontology, and Animal Life*. Albany: State U of New York P, 1999. 1–14.

————. "They Say Animals Can Smell Fear." *Animal Others: On Ethics, Ontology, and Animal Life*. Albany: State U of New York P, 1999. 133–178.

————. "Illicit Crossings." *The Things Themselves: Phenomenology and the Return to the Everyday*. Albany: State U of New York P, 2006. 17–47.

————. "Lost Dog." *The Things Themselves: Phenomenology and the Return to the Everyday*. Albany: State U of New York P, 2006. 48–63.

————. "Rachel Rosenthal Is an Animal." *Mosaic: A Journal for the Interdisciplinary Study of Literature* 39, 4 (December 2006): 1–26.

Stier, Ken, and Emmett Hopkins. "Floating Hog Farms: Industrial Aquaculture Is Spoiling the Aquatic Commons." *The CAFO Reader: The Tragedy of Industrial Animal Factories*. Ed. Daniel Imhoff. Berkeley and Los Angeles: The U of California P and the Foundation for Deep Ecology, 2010. 147–158.

Sunstein, Cass R., and Martha C. Nussbaum, eds. *Animal Rights: Current Debates and New Directions*. Oxford and New York: Oxford UP, 2004.

Talbot, Michael. *Mysticism and the New Physics*. New York: Routledge & Kegan Paul, 1981.

Taylor, Angus. *Animals & Ethics*. Peterborough, Ontario: Broadview Press, 2003.

Thomas, Keith. *Man and the Natural World: Changing Attitudes in England 1500–1800*. New York: Penguin, 1983.

Tietz, Jeff. "Boss Hog: The Rapid Rise of Industrial Swine." *The CAFO Reader: The Tragedy of Industrial Animal Factories*. Ed. Daniel Imhoff.

Berkeley and Los Angeles: The U of California P and the Foundation for Deep Ecology, 2010. 109–124.

Tiffin, Helen. "Foot in Mouth: Animals, Disease, and the Cannibal Complex." *Mosaic: A Journal for the Interdisciplinary Study of Literature* 40, 1 (March, 2007): 11–26.

Twine, Richard. *Animals as Biotechnology: Ethics, Sustainability, and Critical Animal Studies*. London and Washington DC: Earthscan, 2010.

Waldau, Paul. *The Specter of Speciesism: Buddhist and Christian Views of Animals*. Oxford and New York: Oxford UP, 2002.

Waldau, Paul, and Kimberley Patton, eds. *A Communion of Subjects: Animals in Religion, Science, and Ethics*. New York: Columbia UP, 2009.

Wall, Thomas Carl. *Radical Passivity: Levinas, Blanchot, and Agamben*. Albany: State U of New York P, 2009.

Walters, Kerry S., and Lisa Portmess, eds. *Religious Vegetarianism: From Hesiod to the Dalai Lama*. Albany: State U of New York P, 2001.

Webster's Unabridged Dictionary. 2nd edition. New York: Random House, 1998.

Westfall, Richard. *The Construction of Modern Science: Mechanisms and Mechanics*. Cambridge, London, New York: Cambridge UP, 1977.

Williams, Raymond. *Keywords*. London: Fontana, 1983.

Wills, David. *Dorsality: Thinking Back Through Technology and Politics*. Minneapolis: U of Minnesota P, 2008.

———. "Order Catastrophically Unknown." *Mosaic: A Journal for the Interdisciplinary Study of Literature* 44, 4 (December, 2011): 21–41.

Wilmut, Ian, and Roger Highfield. *After Dolly: The Promise and Perils of Human Cloning*. New York and London: Norton, 2006.

Wilson, Elizabeth A. *Psychosomatic: Feminism and the Neurological Body*. Durham: Duke UP, 2004.

Wolch, Jennifer. "Zoöpolis." *Animal Geographies: Place, Politics, and Identity in the Nature-Culture Borderlands*. Eds. Jennifer Wolch and Jody Emel. London: Verso, 1998. 119–138.

Wolch, Jennifer, and Jody Emel. "Preface." *Animal Geographies: Place, Politics, and Identity in the Nature-Culture Borderlands*. Eds. Jennifer Wolch and Jody Emel. London: Verso, 1998. xi–xx.

———. "Witnessing the Animal Movement." *Animal Geographies: Place, Politics, and Identity in the Nature-Culture Borderlands*. Eds. Jennifer Wolch and Jody Emel. London: Verso, 1998. 1–24.

Wolfe, Cary, ed. *Zoontologies: The Question of the Animal*. Minneapolis and London: U of Minnesota P, 2003.

————. *Animal Rites: American Culture, the Discourse of Species, and Posthumanist Theory*. Chicago and London: U of Chicago P, 2003.

————. "Thinking Other-Wise. Cognitive Science, Deconstruction and the (Non)Speaking (Non)Human Subject." *Animal Subjects: An Ethical Reader in a Posthuman World*. Ed. Jodey Castricano. Waterloo: Wilfrid Laurier UP, 2008. 125–143.

————. "Humanist and Posthumanist Antispeciesism." Paola Cavalieri, *The Death of the Animal*. New York: Columbia UP, 2009. 46–58.

Zamir, Tzachi. *Ethics and the Beast: A Speciesist Argument for Animal Liberation*. Princeton and Oxford: Princeton UP, 2007.

Zukav, Gary. *Dancing Wu Li Masters: An Overview of the New Physics*. New York: HarperCollins, 1979.

Index